German Uniforms, Insignia & Equipment 1918-1923

Also by the Author

Uniforms & Equipment of the Imperial German Army 1900-1918 (two volumes)

The Kaiser's Army In Color: Uniforms of the Imperial German Army as Illustrated by Carl Becker 1890-1910

First to the Front: The Aerial Adventures of 1st Lt. Waldo Heinrichs and the 95th Aero Squadron 1917-1918

The Hat in the Ring Gang: The Combat History of the 94th Aero Squadron in World War I

German Uniforms, Insignia & Equipment 1918-1923

• Freikorps • Reichswehr • Vehicles • Weapons

Charles Woolley

Schiffer Military History
Atglen, PA

Acknowledgments

Without the generous sharing of photographs and background information about these times of revolution and change in Germany, this volume could not have been assembled with the depth it enjoys. Craig Brown, German weapons expert extraordinaire and collector of many things to do with Germany's military history, gave me a great deal of help and encouragement. A special thanks to him, one of my more severe critics. Thanks to Art Sylvie who shared many of his wonderful photographs from this period, as did Rick Lundstrom who can identify personalities by the number and placement of award loops on their uniforms, to Richard Bennett a World War I aviation guru who shared photos of some of those who flew with the Freikorps, to David Fletcher of the Tank Museum, Bovington, England, for finding some jewels from that collection, to Jeff Hanson who helped me to find out-of-print books from Germany on the subject, to Peter Harrington, friend and Curator of the Anne S.K. Brown Military Collection at Brown University, who allowed me access to hard to find uniform iconography from that collection, to Ray McGuire who found great Freikorps references for me, and finally to my real pal, Bob Biondi who allows me to wander off into uncharted mine fields of military history with good grace, enthusiasm, and support. Without faithful Nancy, my tough editor with magic fingers, there again would be no book. Thanks, my dear.

Book design by Robert Biondi.

Library of Congress Catalog Number: 2002101182.

Printed in China.
ISBN: 0-7643-1663-X

We are always looking for people to write books on new and related subjects. If you have an idea for a book, please contact us at the address below.

Published by Schiffer Publishing Ltd.
4880 Lower Valley Road
Atglen, PA 19310
Phone: (610) 593-1777
FAX: (610) 593-2002
E-mail: Schifferbk@aol.com.
Visit our web site at: www.schifferbooks.com
Please write for a free catalog.
This book may be purchased from the publisher.
Please include $3.95 postage.
Try your bookstore first.

In Europe, Schiffer books are distributed by:
Bushwood Books
6 Marksbury Ave.
Kew Gardens
Surrey TW9 4JF
England
Phone: 44 (0)208 392-8585
FAX: 44 (0)208 392-9876
E-mail: Bushwd@aol.com.
Free postage in the UK. Europe: air mail at cost.
Try your bookstore first.

Introduction

The intent of this and the volume to follow is to illustrate the insignia, uniforms, and equipment worn and used by military factions in Germany during the years of late 1918 through 1934. These factions consisted of the Imperial Army returned home in defeat, the Provisional Reichswehr, or troops loyal to the Government, the Spartakus, or Communist Revolutionary army, the red Volksmarine Division, the Freikorps, the Reichswehr after 1919, the Reichsmarine, the Stahlhelm, the S.A., the Hitler Youth, and other para-military uniformed units.

Post-World War I Germany went through a series of wrenching events during this sixteen-year period, each had its effect on military dress and equipment. The Naval mutinies, the first in 1917 and the second in November 1918, were the beginnings of the Bolshevik attempt toward revolution. Communist agitators like Rosa Luxemburg and Karl Liebknect, both later murdered by the Freikorps, began to fan the fires of revolution in earnest. After the Armistice, the People's Naval Division traveled from Kiel to Berlin and by force took over the Imperial Palace and Stables (Schloss und Marstall). A pitched battle for the Palace took place on the morning of Christmas Eve 1918 between Garde units of the old Imperial army and the rabble occupying the buildings which ended in a short twenty-minute truce. Members of the Volksmarine Division and other Spartakus leaders during that twenty minutes jumped into open cars and were driven through the streets of Berlin shouting, "The Republic is in danger! Imperial officers are seizing the Schloss. There is a monarchist counter-revolution. Come to the Schloss and save the Republic!" A large hostile crowd gathered quickly, more firing broke out, and the confused Garde troops not wishing to fire on the crowd, broke off the engagement and retired to their barracks, and the Communists continued to occupy the Palace. After Christmas, Gustave Noske was called from Kiel to Berlin by Chancellor Ebert and his government, to become the Supreme Commander of the Forces (Oberfehlshaber) with the task of reshaping the Army. On January 4, 1919, Noske and Ebert traveled to the outskirts of Berlin to a meeting with General von Maercker. At Camp Zossen they reviewed the first well-disciplined and well-equipped Freikorps unit called Maercker's Volunteer Rifles. This was the beginning of the Freikorps movement and with it, the stabilization, in part, of the Army.

Over the next few years the ranks of the Freikorps grew as required by the degree of fighting on various fronts. They were desperately needed to reclaim Berlin and Munich from Spartakus control, to meet the threats of Russian, Polish, and Revolutionary Germans in the Baltic, to overcome Revolutionary actions in the Ruhr, as well as other major cities. These all were campaigns fought by the varied units, both large and small, and with the exception of the Baltic debacle, were usually successful.

During this time the Reichswehr was in the throes of reconstruction. The hated Versailles Treaty put limitations on the size of the Army, 100,000 troops, as well as armament. No tanks, no aircraft, in fact weapons of an offensive nature were banned or dramatically curtailed. Stocks of the Imperial uniforms of the M 1910 and M 1915 regulations continued to be issued, but designs for new uniforms began in early 1919 reaching some units very quickly. The Freikorps troops were issued the leftover wartime material, while many individuals continued to use uniforms issued to them before the cessation of hostilities. The pride in their individual volunteer units caused a major movement in the design and creation of collar, sleeve and breast badges to be worn by each as identification. Literally hundreds of these badges were designed and worn during this short period, many defying positive identification. The Freikorps was dealt a hard to swallow blow by the dictates of the Versailles Treaty demanding the disbanding of these volunteer units. General Hans von Seeckt was given the reins to rebuild the Reichswehr with the best available manpower. He was less than enthusiastic about the methods and ways of the freebooter-style of Freikorps units and, therefore, refused to employ many of them in his new army. General von Epp's men although were absorbed into the Reichswehr, but the Ehrhardt Brigade, Rossbach's Freikorps and many other of the more radical units, were not acceptable to von Seeckt and were forced underground to emerge later under the National Socialists banner. Before the Freikorps could be disbanded,

General Lüttwitz and Wolfgang Kapp, in an attempt to save the Freikorps, quickly planned a putsch to overthrow the government in power and install one with pro-Imperial military and anti-Versailles Treaty leanings. The putsch failed and the Freikorps as it was known went with it. The "alte kampfers" formed the Stahlhelm with its unique uniforms, and many of those Freikorpskampfers not absorbed into the Reichswehr were recruited into the ranks of the National Socialist or Nazi party and formed SA and SS units to further that cause. The SA adopted an array of different kepi colors, rank tabs, and epaulets to adorn their brown shirts and dress tunics, the SS in black tunics and brown shirts were more somber in their dress. Uniforms and distinctive insignia abounded by 1933, including those of the Stahlhelm, Hitler Jugend, Deutscher Luftport-Verband, Landspolizie, Arbeitschienstwilliger, Railway Police, Ehrhardt Brigade, Postal Workers, Border Guards, Forestry, Feldjäger Korps, Marine-werstpolizei, and finally the National Reichswehr and Reichsmarine.

The uniforms seen in this volume are first those with wartime origins, M 1910 tunics, M 1915 Bluse, M 1915 Friedens tunics, then 1919 "Regulation" tunics, 1919-1920 Waffenrocks, Dienstmützen, Feldmützen, Kratzchen and Stahlhelm M 16, M 18, M 18 with cutouts. Weapons used were those of wartime manufacture and origin. Rifles and carbines Gew. 98, Kar. 98a, a few Gew. 88s and Kar. 88s, P. 08s, M 96 Mausers, MG 08s, MG 08/15s and MP 18s. In the heavier calibers, light Minenwerfers, heavy Minenwerfers, 7.7 cm, M 96 n/A FK and 10.5 cm LFK were in operation. Wix Flammenwerfers saw use in the Berlin fighting, as were a wonderful assortment of tanks and armored cars used in Berlin, Munich and finally the Ruhr.

The illustrations, both black and white, plus color, are from the 1918 to 1923 time period and whenever possible, unpublished photos have been used. This book is not intended to include an illustration of every Freikorps insignia or badge, or offer a list of the names of every unit, but does try to give a good cross-section of how things actually looked during these rather chaotic times. An effort has been made to include images of the Communist Revolutionary forces which are not as numerous as those of the Reichswehr and the Freikorps. The shadowy and disorganized style of those who fought under "Die Rote Fahne" did not lend itself to be heavily photographed, but the pictures included will give the reader a good notion as to how they appeared.

OPPOSITE
This oil painting by Gerhard Grossman entitled *Matrosenaufstand 1918* illustrates most graphically the mood of the Naval mutinies of late 1918.

German Uniforms, Insignia & Equipment 1918-1923

Left: A Freikorpskampfer readied for action. Dressed in an M 1915 Bluse with shoulder straps from the Alte Armee, plus a ribbon for the Iron Cross Second Class. His equipment consists of a Kar. 98a, M 1909 Patronentasche cartridge boxes and Bz. 15 grenades. Of particular interest is the M 18 Stahlhelm decorated with a Totenkopf, as well as other designs surrounding and below the ventilation lugs. What the complete design entailed and its meaning remain a mystery. Behind our man is a 7.7 cm FK 96 n/A field gun.

OPPOSITE

Top: Troops of the Garde Jäger Bataillon recently returned from the then still Western front, form up in the Pariser Platz, Berlin on 11 December 1918. They wear the M 1915 Bluse with Garde collar Litzen, a mixture of Ersatz and leather Tschako's and are equipped with Gew. 98s, 1909 Patronentasche supported by the M 1909 Koppel, or service belt. Only one man is still wearing stiefel, or marching boots, while the others are in legwraps, or Wickel-gamaschen, and laced boots, or Schnürschuh. In the background steel-helmeted troops are forming with a tri-colored National flag of black, white and red.

Bottom: This is a group photo of Republican, or Provisional Reichswehr troops, as indicated by the absence of any Freikorps insignia. All are equipped with Kar. 98a's with one exception who is equipped with a Gew. 98. M 1909 Patronentasche and M 1918 Stalhelm, Bz. 15 grenades, a P. 08 and an MG 08/15 are seen in this picture taken in Freiberg by photographer Oscar Hertel.

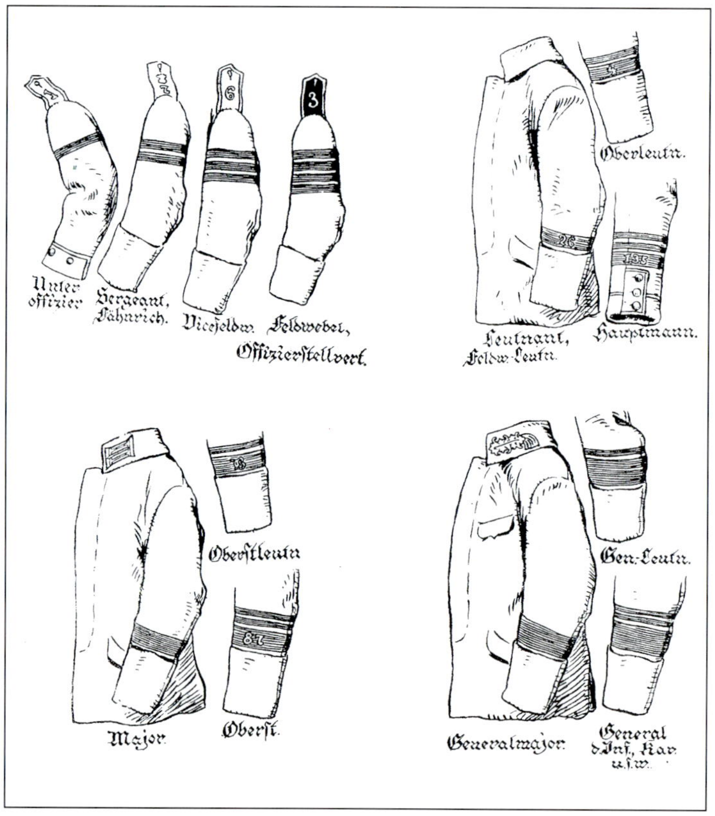
Unter offizier
Sergeant, Fähnrich.
Vicefeldw.
Feldwebel, Offizierstellvert.
Oberleutn.
Leutnant, Feldw.-Leutn.
Hauptmann.
Oberstleutn.
Major
Oberst.
Gen.-Leutn.
Generalmajor.
General d. Inf., Kav. u.s.w.

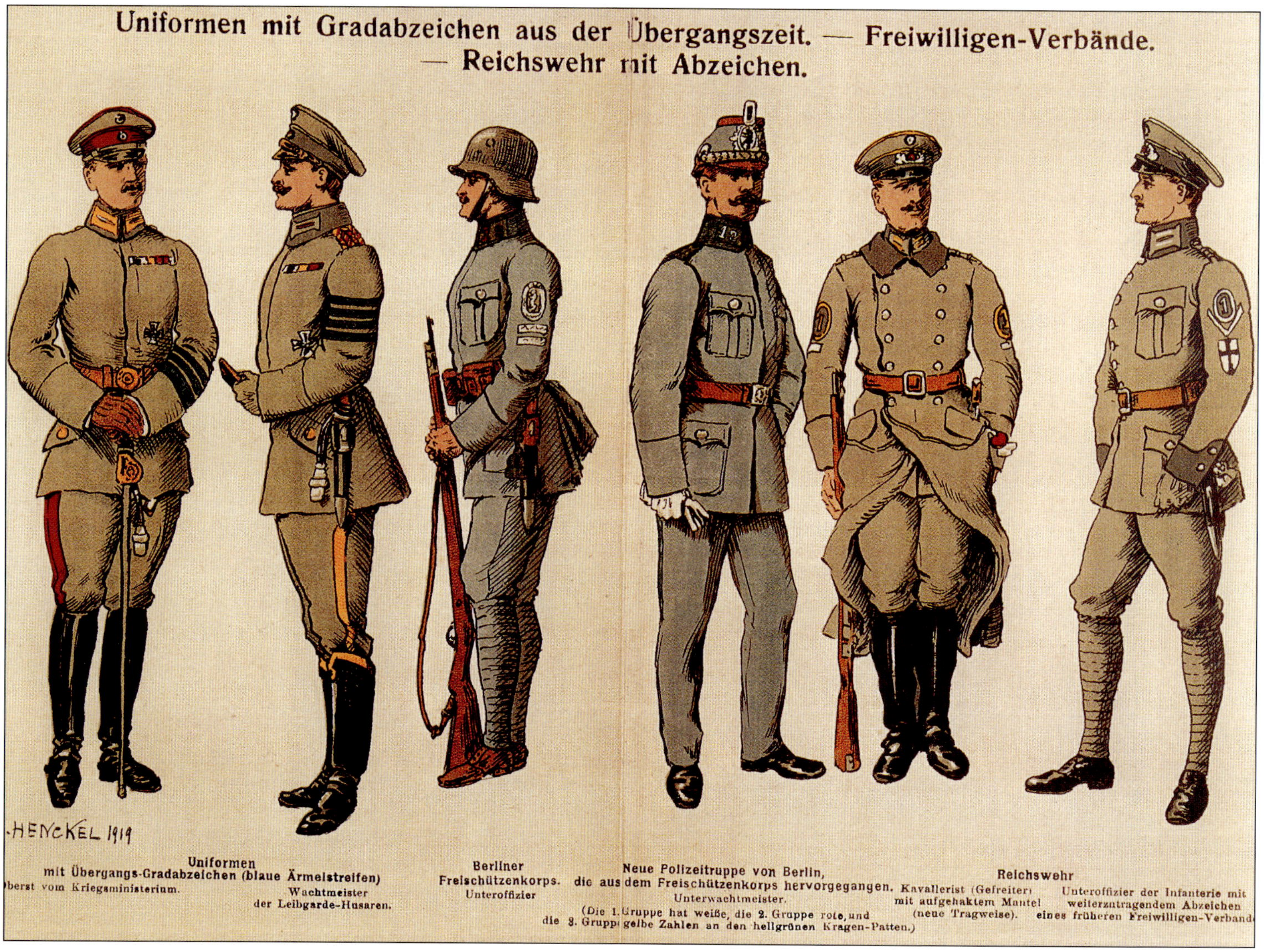
Uniformen mit Gradabzeichen aus der Übergangszeit. — Freiwilligen-Verbände. — Reichswehr mit Abzeichen.
HENCKEL 1919
Uniformen mit Übergangs-Gradabzeichen (blaue Ärmelstreifen)
berst vom Kriegsministerium.
Wachtmeister der Leibgarde-Husaren.
Berliner Freischützenkorps. Unteroffizier
Neue Polizeitruppe von Berlin, die aus dem Freischützenkorps hervorgegangen. Unterwachtmeister.
(Die 1. Gruppe hat weiße, die 2. Gruppe rote, und die 3. Gruppe gelbe Zahlen an den hellgrünen Kragen-Patten.)
Reichswehr
Kavallerist (Gefreiter) mit aufgehaktem Mantel (neue Tragweise).
Unteroffizier der Infanterie mit weiterzutragendem Abzeichen eines früheren Freiwilligen-Verband

OPPOSITE
Top left: The sleeve rank insignia of the Friedensheer, or peacetime, army dictated by the regulations of 19 January 1919.

Top right: An illustration by the German uniform historian Paul Pietsch shows the early Reichswehr sleeve insignia as worn on the uniform in 1919. This and the previous illustration appeared in Issue Number 4, 1919, of the rare *Archiv für Waffen und Uniformkunde*.

OPPOSITE
Bottom, three figures left: Uniforms in transition. (Ubergangszeit) for Volunteers and the Reichswehr, 1919. Figure 1 depicts a Colonel of the War Ministry clearly showing the dark blue sleeve rank insignia. Figure 2 is a Wachtmeister of the Leibgarde Husaren no longer wearing the Husar Attila, but retaining Husar boots and yellow piping on the breeches, as well as Husar-style shoulder straps. Figure three is an NCO of the Berlin Freischützenkorps. During the riots and unrest in Berlin, the Freischützenkorps was the voluntary military arm of the police. The badge on his sleeve shows the Berlin bear symbol worn over two rank stripes.

Three figures right: The second phase of the uniform transition, 1919. Figure 1 shows the uniform of a Unterwachtmeister of the new Berlin Police Force. Figure 2 depicts a Gefreiter of the 1st Cavalry regiment in the Mantel, or overcoat. The yellow piping of the cavalry is in evidence on the Dienstmütze and the oval sleeve unit designation (Armspiegel). The NCO shown in Figure 3 has white piping on his cap and unit insignia indicating he is in the Infantry. He also wears below his rank insignia the insignia of the Freiwilligen-Verband v. Kretschmann. The new uniform design of May 1919 with its exposed buttons, green collar with Litzen, green cuffs, four pockets, and narrow shoulder straps is clear in this illustration.

Above: A 1919 photo of Hermann Ritter von Mann Edler von Tiechler, shown as an Oberleutnant while serving as the Deputy Company C.O. of Provisional Reichswehr Infantry Regiment 48. His tunic is fly-fronted with exposed upper pockets, collar Litzen, narrow shoulder cording and the rank insignia on the cuff. The silver-pierced Wound Badge is interesting as is the dress bayonet worn on a claw-buckled leather belt. Von Tiechler was a P.O.W. in England from 1916-1919 and later became a Generalleutnant of the Luftwaffe in 1940.

Left: An NCO of the Reichswehr cavalry wearing what appears to be a slightly modified 1915 style Feldrock with exposed buttons, slash lower pockets, narrow shoulder straps, Litzen on collar, and regimental Armspiegel. The insignia on his cap consists of two cocards, the lower surrounded by a small Reichswehr-style wreath, while the cap itself has a dented crown affected by many NCOs and officers.

Left: Another example of transitional uniforms and insignia. The cuffs and collar are in green on the "Regulation" uniform, which has a hidden button, or fly-front, with exposed upper pockets and slash lower pockets. The single rank bar of a Gefreiter is worn over a chevron of a type worn by Freikorps troops. His Dienstmütze has the popular "pinched" crown and bears the Reichskokard with the Reichswehr symbol in a wreath. An Iron Cross Second Class ribbon and a black Wound Badge are worn, as well as a dress bayonet without a knot, or Troddel.

OPPOSITE
A Reichswehr cavalryman dressed in an M 1915 Bluse with the shoulder boards removed, Litzen added to the collar and a unit designation oval Armspiegel. He wears a white dress belt with sword slings for the Mannschaftssäbel hung behind him on the saddle and cavalry hobnailed boots, or Reitstiefel of World War I style. His Dienstmütze carries both cocards with the lower surrounded by a wreath of oak leaves.

Above: Members of the Communist Revolutionary "Workers Batallion" march in the Schlossplatz, or castle square, Berlin, December 1918. The dress varies from Naval sailors and Army soldiers in full uniform to "Arbieters" or workers in civilian clothes of all types. The arms universally in this picture are Gew. 98s, but ammunition is not in evidence.

OPPOSITE

Top: The opposition to the Reds, the Freikorps and Government troops streetfighting in Berlin. An interesting mixture of uniforms and equipment can be seen in this photograph: M 1915 Bluse and M 1910 tunics, cavalry boots and spurs, empty bayonet frogs, leg-wraps and Schnürschuh M 1914 laced boots, visor caps, steel helmets, Kar. 98a's, and the ever present Bz. 15 grenades hung from the waistbelt.

Bottom: After the fighting of December 23-24, the Communist Volksmarine Division buries their dead. The funeral cortège leaves the Kaiser's Palace. December 29, 1918. All the pallbearers wear Naval uniforms of the Volksmarine Division.

KAUFHAUS FÜR DAMENPUTZ
ADOLF COHN Jr.
Herren-
Bekleidung
Bluhm

Arbeiter-
helft

OPPOSITE

Top: Hot food is served to loyal Government troops from a "Gulaschkanone" (Field Kitchen) during the Berlin streetfighting, March 1919. The men are equipped with Gew. 98 rifles, Ersatz and 98/05 bayonets, M 16 and M 18 helmets, M 1915 Bluse, feldgrauer Mantels M 1915, side-laced and M 1914 Schnürschuh (3/4-laced marching shoes) with leg-wraps (Wickel-gamaschen) and the ever present hand grenades Stielhandgrenate Bz. 15. The NCO on the right has his visor cap stuffed into his belt.

Bottom: The Red Volksmarine Division occupies the Kaiser's Palace and stables, December 1918. A mixture of Naval and Army uniforms are present with the majority being Naval. In this picture these men are still wearing the red, white and black cocards on their caps. As the revolution progressed, all cocards were discarded. An array of weapons can be seen: MG 08/15s, MG 08s, Gew. 98s, Kar. 98a's, and at least one example of a holstered P.08. In the center of the courtyard rests what looks to be a 175 mm Minenwerfer. The sailor on the far right has his Kar. 98a slung muzzle down, as was the habit of many of the revolutionaries.

Gustav Freiheir von Perfall photographed while serving as a staff officer with the 21st Provisional Reichswehr Brigade and wearing a uniform which has World War I Bavarian buttons, but the 1919 style shoulder cords. The lion head on the black diamond indicates his service in the Freikorps von Epp and the Armspiegel bearing "21", his assignment to that unit. The order is the Bulgarian Military Order 3rd Class with War Decoration, the pin-back medals are: Iron Cross First Class, Brunswick Ernst August Cross First Class, Austro-Hungarian Franz Joseph Order-Officer Class with War Decoration, 1918 black Wound Badge. The ribbon bar consists of Bavarian, Prussian, Brunswick, Württemberg, Bulgarian and Austro-Hungarian awards. During World War I he served as a Rittmeister in the Bavarian 1.Schwere Reiter Regiment and a Flügeladjutant to the King of Bavaria. He was promoted to Generalleutnant in 1939.

In the fall of 1919, Paul Pietsch, the well-known German Army uniform historian, produced a short monograph entitled *Die deutsche Reichswehr* which was published for the Archiv für Waffen und Uniformkunde by F. Rascher in Frankfurt. This publication served as an explanation to the new regulations for uniforms set down on 5 May 1919. The four black and white illustrations which follow are Pietsch's illustrations from the publication as they appeared.

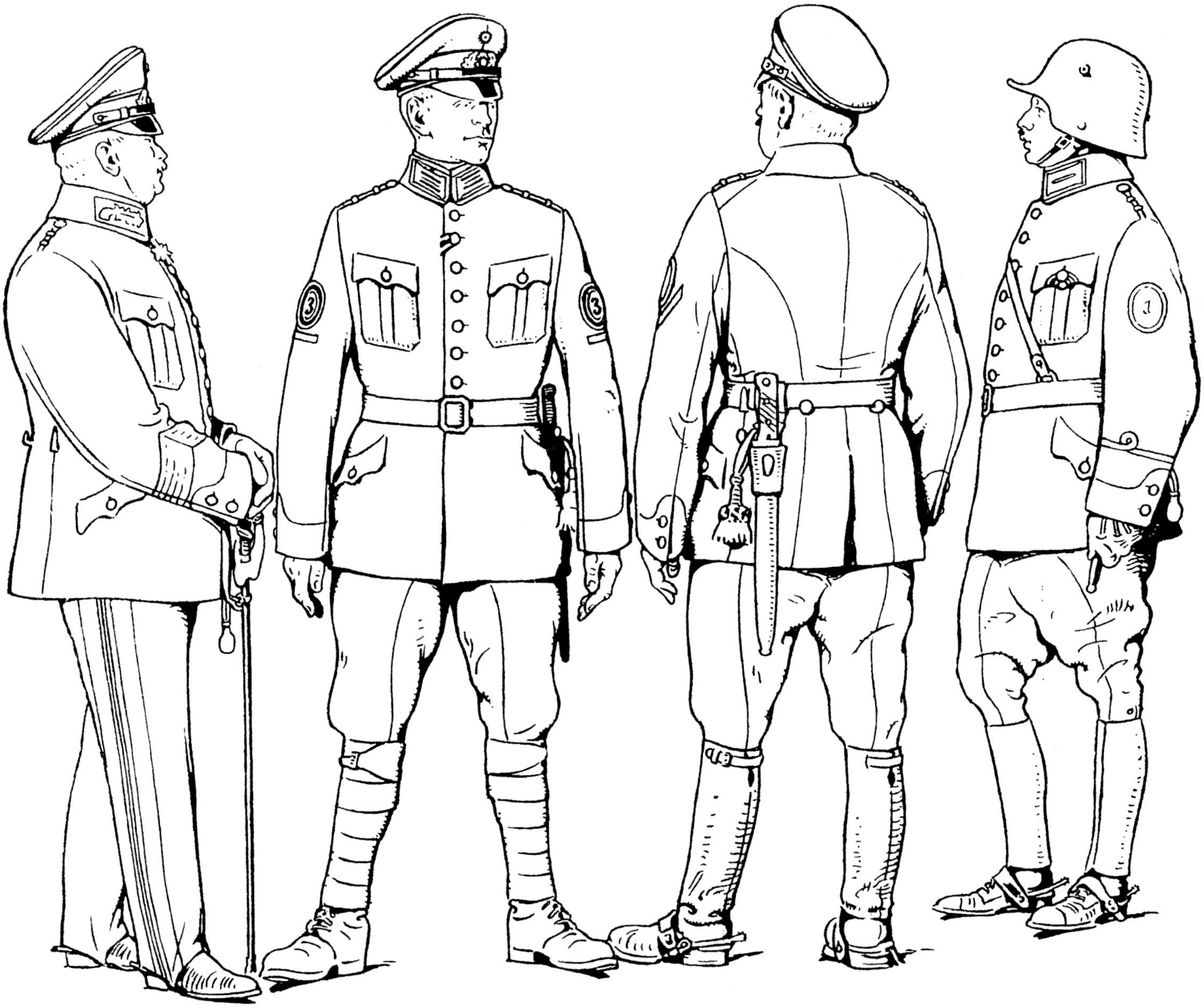

General. Fusstruppen. Berittene. Leutnant.

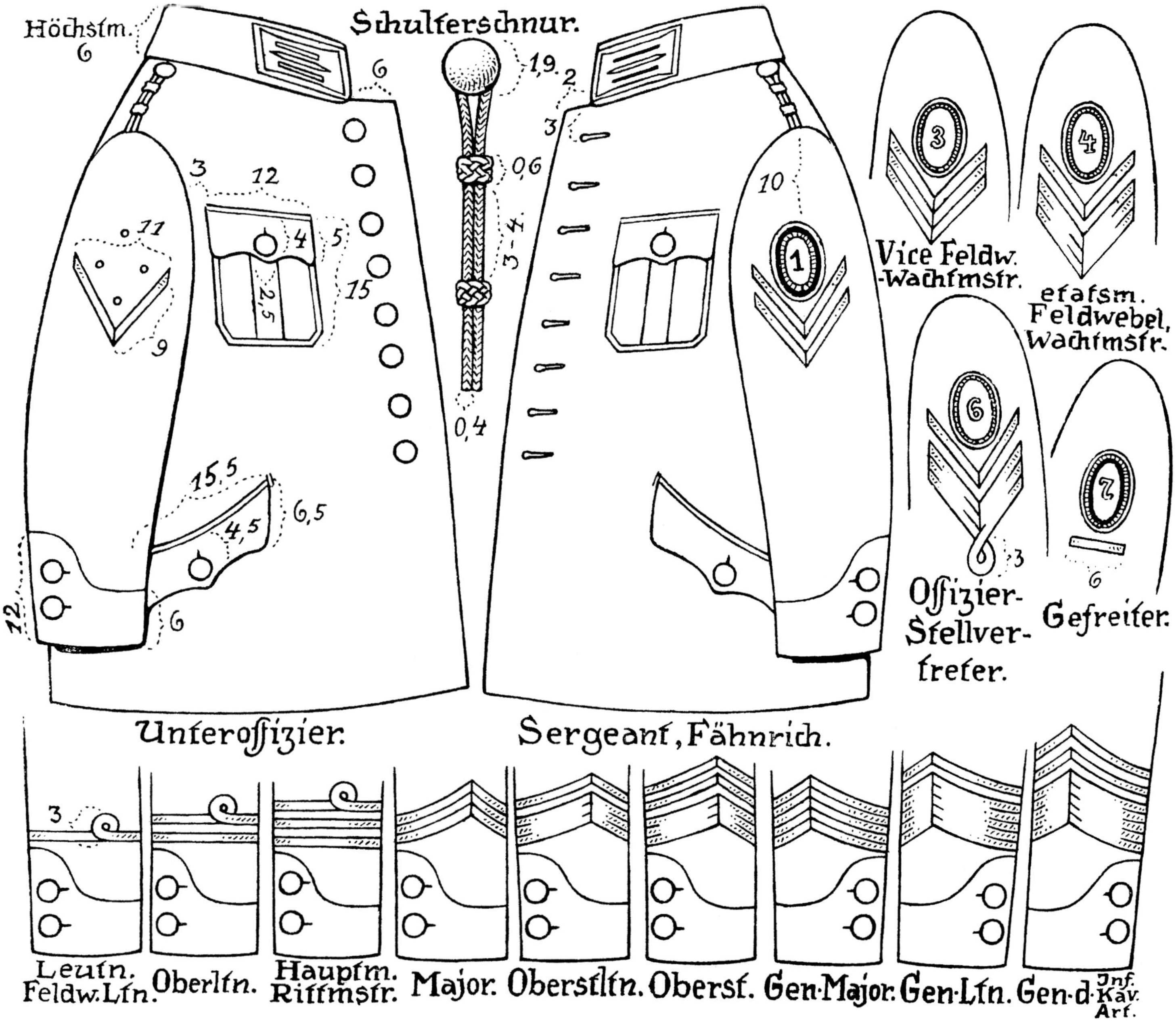

Höchstm. 6
Schulterschnur.
Unteroffizier.
Sergeant, Fähnrich.
Vice Feldw. -Wachtmstr.
etatsm. Feldwebel, Wachtmstr.
Offizier-Stellvertreter.
Gefreiter.
Leutn. Feldw. Ltn.
Oberltn.
Hauptm. Rittmstr.
Major.
Oberstltn.
Oberst.
Gen-Major.
Gen-Ltn.
Gen-d- Inf. Kav. Art.

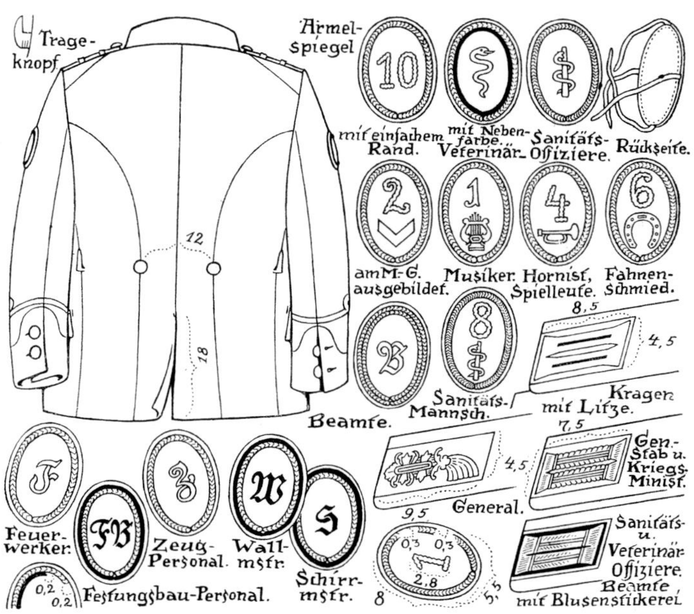

Trage-knopf
Ärmel-spiegel
mit einfachem Rand.
mit Nebenfarbe. Veterinär.
Sanitäts-Offiziere.
Rückseite.
am M-G. ausgebildet.
Musiker.
Hornist, Spielleute.
Fahnen-Schmied.
Beamte.
Sanitäts-Mannsch.
Kragen mit Litze.
Gen-Stab u. Kriegs-Minist.
General.
Sanitäts-u. Veterinär-Offiziere Beamte mit Blusenstickerei.
Feuer-werker.
Zeug-Personal.
Wall-mstr.
Schirr-mstr.
Festungsbau-Personal.

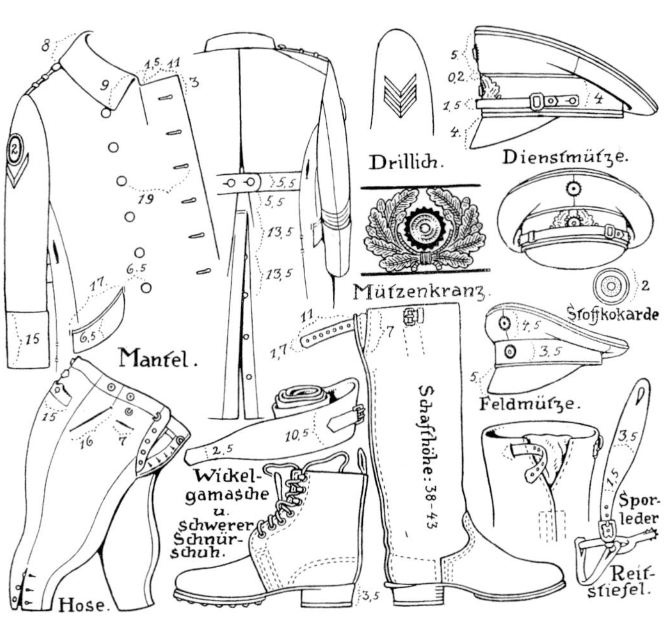

Drillich.
Dienstmütze.
Mützenkranz.
Stoffkokarde
Mantel.
Feldmütze.
Schafthöhe: 38-43
Wickel-gamasche u. schwerer Schnür-schuh.
Sporleder
Hose.
Reit-stiefel.

About the same time Pietsch was illustrating the new uniforms and insignia in Archiv für Waffen und Uniformkunde, Moritz Ruhl, publisher in Leipzig, published *Die Deutsche Reichswehr* with seven folding color plates of the uniforms and insignia, included here. Ruhl had been publishing these quasi-official uniform books with folding color plates since the 1880s and continued, changing the format slightly, into the 1930s. The regulations for these new 1919 uniforms were the base from which those of the 1920s and 1930s grew, and the changes will be shown in the order in which they occurred. The captions in these illustrations have been left in German to keep a continuity of terms of these and future plates in this volume.

Die neue Deutsche Reichswehr.
Infanterie
Soldat
Kavallerie
Gefreiter
Leichte Artillerie
Unteroffizier.
Train
Sergeant.
Pionier
Vizefeldwebel.

Die neue Deutsche Reichswehr.
Schützen.
Feldwebel
Sanitätsoffizier
mit Leutnantsrang.
Jäger
Oberleutnant.
Kavallerie
Major.
Generalstab
Hauptmann.

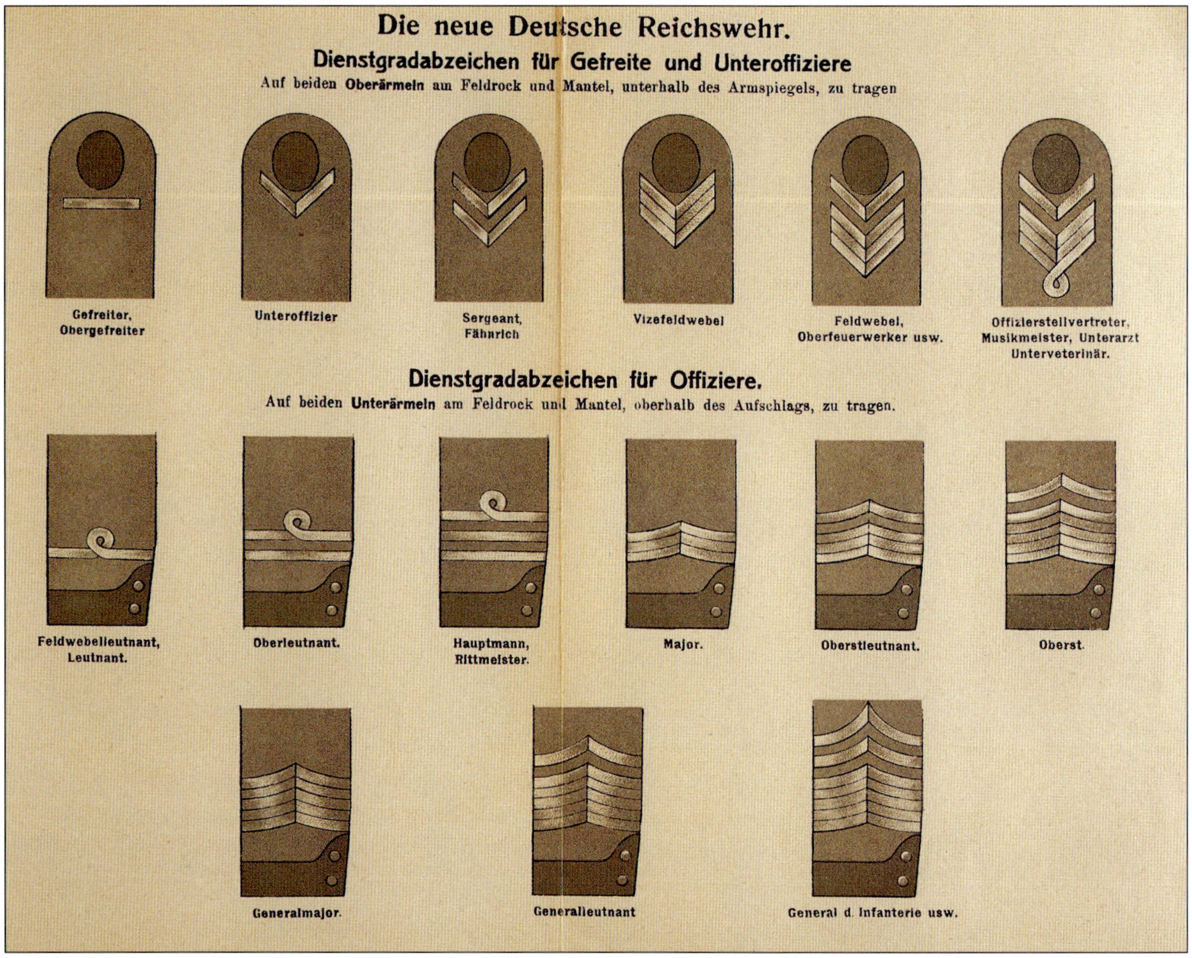
Die neue Deutsche Reichswehr.
Dienstgradabzeichen für Gefreite und Unteroffiziere
Auf beiden Oberärmeln am Feldrock und Mantel, unterhalb des Armspiegels, zu tragen
Gefreiter, Obergefreiter
Unteroffizier
Sergeant, Fähnrich
Vizefeldwebel
Feldwebel, Oberfeuerwerker usw.
Offizierstellvertreter, Musikmeister, Unterarzt Unterveterinär.
Dienstgradabzeichen für Offiziere.
Auf beiden Unterärmeln am Feldrock und Mantel, oberhalb des Aufschlags, zu tragen.
Feldwebelleutnant, Leutnant.
Oberleutnant.
Hauptmann, Rittmeister.
Major.
Oberstleutnant.
Oberst.
Generalmajor.
Generalleutnant
General d. Infanterie usw.

Die neue Deutsche Reichswehr.
Armspiegel zur Kennzeichnung der Waffengattung und des Truppenteils.
Auf beiden Oberarmen am Feldrock und am Mantel anzulegen.
Infanterie
Schützen
Jäger
Maschinengewehr-Abtlgn.
Maschinengewehr-Scharfschützen-Abtlgn.
Gebirgs-Maschinengewehr-Abtlgn.
Kavallerie.
Leichte Artillerie.
Schwere Artillerie.
Flakformationen.
Gebirgsartillerie.
Minenwerfer.
Artillerie-Messtrupps.
Artillerie-Parks.
Train.
Traindepots.
Pioniere.
Ingenieurkorps.
Fliegertruppen.
Luftschiffertruppen.
Eisenbahntruppen
Kampfwagentruppen
Kraftfahrtruppen.
Nachrichtentruppen

Die neue Deutsche Reichswehr.
Armspiegel zur Kennzeichnung der Waffengattung und des Truppenteils.
Auf beiden Oberarmen am Feldrock und am Mantel anzulegen.
Vermessungsabteilungen
Sanitätsformationen
Sanitätsoffiziere, Unterärzte
Veterinäroffiziere, Unterveterinäre,
Tierblutuntersuchungsstellen.
Bekleidungsämter.
Versorgungsämter.
Landwehr-Inspektionen, Bezirks-Kommandos
Zeugoffiziere, Zeugfeldwebel
Feuerwerksoffiziere, Feuerwerker
Festungsbau-Offiziere u. -Feldwebel
Wallmeister.
Beamte
Militärintendanturen
Militärjustizverwaltung
Zahlmeister Militärkassenwesen.
Waffenmeister.
Apotheker
Zahnärzte
Abzeichen in den Armspiegeln für besondere Dienststellungen.
Spielleute. Bataillonstamboure.
Hoboisten, Musikmeister
Sanitätsmannschaften u. Sanitätsunteroffiziere.
Fahnenschmiede
Abzeichen der Befähigung zum Bedienen eines Maschinengewehrs.

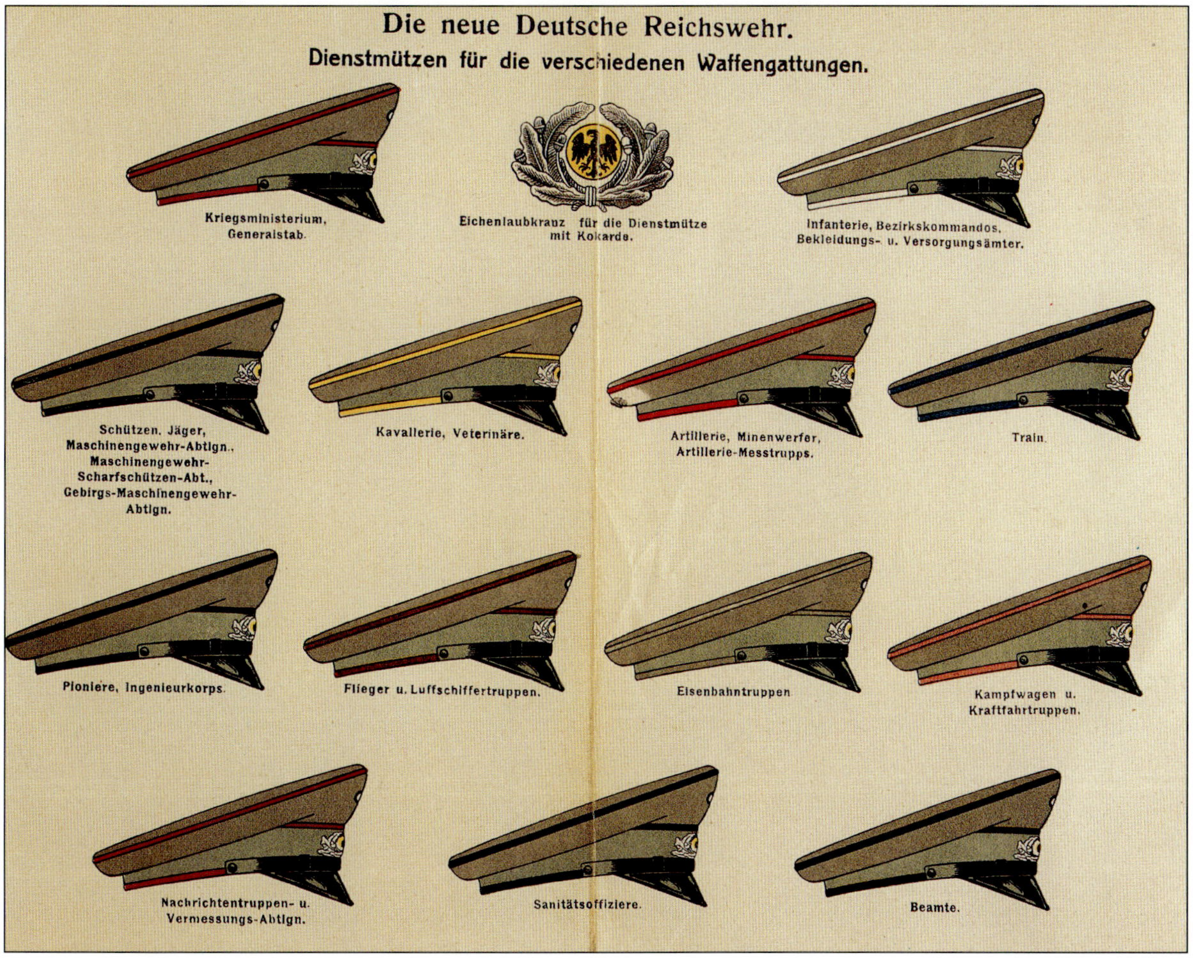
Die neue Deutsche Reichswehr.
Dienstmützen für die verschiedenen Waffengattungen.
Kriegsministerium, Generalstab.
Eichenlaubkranz für die Dienstmütze mit Kokarde.
Infanterie, Bezirkskommandos, Bekleidungs- u. Versorgungsämter.
Schützen, Jäger, Maschinengewehr-Abtlgn., Maschinengewehr-Scharfschützen-Abt., Gebirgs-Maschinengewehr-Abtlgn.
Kavallerie, Veterinäre.
Artillerie, Minenwerfer, Artillerie-Messtrupps.
Train.
Pioniere, Ingenieurkorps.
Flieger u. Luftschiffertruppen.
Eisenbahntruppen
Kampfwagen u. Kraftfahrtruppen.
Nachrichtentruppen- u. Vermessungs-Abtlgn.
Sanitätsoffiziere.
Beamte.

Alte Frontsoldaten
Minenwerfer
Maschinengewehrschützen
Freiwillige
gediente und ungediente
Schützt Euer
bedrohtes
VATERLAND
Meldet Euch
beim Freiwilligen-(F.)
Garde-Gren.-Bataillon
SPANDAU
Neue Kaserne Schmidt-Knobelsdorfstr.
(am Potsdamer Platz)
R.Pohle 1919
Nauck & Hartmann Berlin C.19
Bedingungen: Mobile Löhnung und 5 Mark tägliche Zulage, mobile Verpflegung, Unterbringung und Bekleidung
Militärbekleidung und Militärpapiere, soweit vorhanden, mitbringen!
Das Vaterland ist in Gefahr! Eile tut not!
von Negelein, Major und Batls.-Kommandeur.

Gipkens
Schützt Eure Heimat!
Tretet bei dem Freikorps
Hülsen
Centralwerbestelle
Charlottenburg
Oranien Str. 13-15

OPPOSITE
As the Revolutionary movement swelled, it became quickly apparent that for the Army of the Government, help was needed. The Freikorps concept was born, and the following are two examples of recruiting posters which were posted in major cities.

Right: Colonel Franz Ritter von Epp, Commander of the Epp Brigade, one of the strongest Freikorps units in Germany. Near the end of 1919, Epp and his Brigade entered the official Republican Army, or Reichswehr, but continued to call themselves Freikorps von Epp. The von Epp sleeve insignia is prominent on his uniform, as is the Pour-le-Mèrite at his neck.

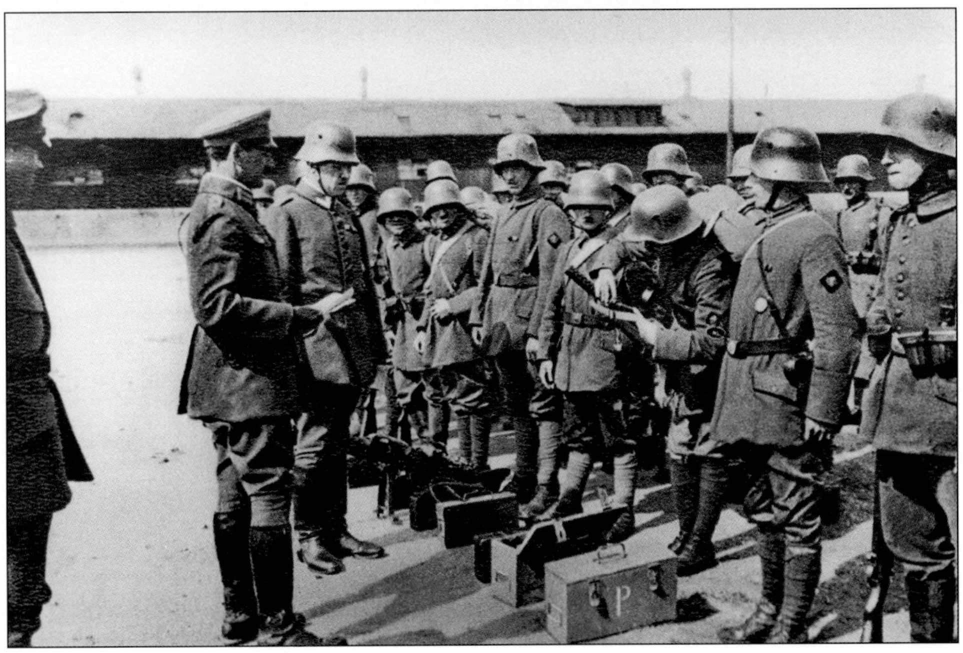

Right: Colonel von Epp inspecting his troops. Of particular interest in this photo is the man third from the right. He is demonstrating the loading of a snail drum magazine into an MP 18 submachine gun. To his right is a machine gun squad with an MG 08/15 and ammunition boxes at their feet. The open box next to the man with the MP 18 in all probability contains extra snail drums with ammunition. The uniform most in evidence is the M 1915 Bluse with the man nearest the camera in an M 1910 tunic. Marching boots and leg-wraps complete their outfits.

FLORES

OPPOSITE - Top: Munich was the scene of heavy streetfighting in March 1919 and this picture shows Freikorps troops patrolling, while civilians seemingly went about their business. The helmeted Husar wears his regimental Attila, leg-wraps, and shined laced boots, is armed with a Kar. 98a, spare ammunition in a gray cloth bandolier "Patronengurtel", and two Bz. 15 grenades. His companion is similarly armed wearing a Dienstmütze, M 1915 Mantel and loose cuffed trousers with laced high shoes. **Bottom**: Bavarians from Jäger Bataillon Nr. 1 man an MG 08 by a storefront in downtown Munich, March 1919. They wear the Bavarian M 1916 Bluse with the Bavarian blue and white borte, or tape, on the collars, with the exception of the man on the left. The man on the far right wears a Red Cross armband and carries the medical kit on his belt. What appear to be bullet holes are in the plate glass window behind them.

Above: This photo postcard was taken during the streetfighting against Revolutionary troops in Munich, March 1919. A weapon used in Munich and Berlin was the Wix M 17 Flammenwerfer, two examples of which are shown here. The crew wears M 1915 Bluse with Litzen on the collars and dark shoulder straps, M 18 cutout helmets, Kar. 98a's, and Bz. 15 grenades. On the left two men wear the M 1915 Mantels, also with dark shoulder straps.

Below: Another view of the same action in close-up. The M 18 cutout helmets and the uniform details can be seen, as well as the nozzle of the M 17 Flammenwerfer.

Left: A recently equipped Freikorpskampfer dressed in what would appear to be a brand new M 1915 Bluse, M 1908 Patronentasche, M 16 helmet, Sturmgepäck M 16, M 17 Infanteriestiefel (Infantry boots), the sleeve insignia of Freikorps von Epp, and is armed with a Kar. 98a.

OPPOSITE
A Hoffmann photo of Oberst von Epp leader of Freikorps von Epp. Hoffmann became Hitler's private photographer and official Nazi party photographer. Von Epp joined the Nazi party at an early point in its history and was one of the first deputies to the Reichstag. Hitler appointed him the Reichstatthalter of Bavaria (Governor of Bavaria).

OBERST EPP

MÜNCHEN
MAI. 1919.

OPPOSITE
Top: The streets of Munich were patrolled and guarded by the Freikorps von Epp. In this photo we can see an MG 08 in the background, helmets at the ready nearby, a 7.7 cm FK 96 n/A field gun and limber, Kar. 98a's and P 08s in holsters which make up the weaponry. These troops are wearing both cocards on their service caps and Mütze. May must have been a cool month as three men are in overcoats.

Bottom: This photo includes Gustav Noske, Reichminister of Defense, the tall man wearing glasses in civilian clothes and von Epp, looking at the camera. Other Reichswehr officers join him to review von Epp's Freikorps troops when they became part of the Reichswehr. The ceremony was in Munich.

Above: Streetfighting in March 1920 involved armored cars, Flammenwerfers, and troops wearing visorless helmets. The armored car is an Ehrhardt E-V/4, and bears the Totenkopf insignia used by the Freiwilligen Panzer-Kraftwagen-Abteilung and other armored units of the Army Kokampf during and after the revolution. With modifications this insignia from early cavalry days was used by Nazi Wehrmacht panzer forces as well as the SS. No unit insignia can be observed on the troops which are wearing the 1919 style Reichswehr tunics.

Above: The Red Guard in Munich in front of the central railroad station. A mixture of Naval and Army uniforms is in evidence. The national cocard has been removed from their assorted caps, weapons include Gew. 98s, a Reichsrevolver, ammunition bandoleers and hand grenades. None of these revolutionaries seem to be pleased to have their picture taken.

Below: This photo postcard is titled "Tausendmann-Kaserne after the battle with the Spartakus on the night of 27-29 January 1919. Corridor of the Kaserne." The photo was taken by Wilhelmshaven photographer Kloppmann and shows loyalist Naval forces who expelled the red Spartakus revolutionaries from the Kaserne. Gew. 98s and a 1906 Naval Model 6-inch barreled Luger are the only weapons visible, although several machine gun ammo boxes are seen on the floor indicating something more than sidearms were used in the fight.

The Government troops during the 8 March 1919 streetfighting in Berlin fire a 10.5 cm light field howitzer at an unseen target. The location of the gun was the Alexanderplatz. The man giving the firing order is armed with a P. 08 in its holster and others are armed with Kar. 98a's. The crew all wear soft caps which indicates no one was shooting back.

Above: Five members of Freischützen-Regiment Berlin are photographed near their Kaserne on 15 May 1919. The man on the left, Gefreiter Gütsche, wears the single rank stripe above the arm badge of the unit which is the Berlin bear enclosed in a crowned wreath. Their collar tabs are marked with a small "3" and a larger "2", the tunics having four pockets which are stitched on the outer part of the tunic, the buttons being enclosed with a fly-front and there are no epaulets or straps or buttons to affix them. The other men with Gütsche are ranked as volunteers and are named left to right: Adam, Büll, Schüchardt, and Romahn.

OPPOSITE

Top: Revolutionaries in Munich pose with their flag stating they are the Rote Garde of the Communist Soviet Republic, and of most interest are the words "Luftschiffer Abteilung." An airship presence by the Rote Garde is an unknown factor. The men in the center surround an MG 08, while others hold Gew. 98s and Kar. 98a's. Their uniforms are varied, and interestingly, the two men prone in the front row wear fleece-lined Pelzmantels M 1915, which were designed for and issued to flying troops, one of which bears a red armband. Also in the picture are both male and female civilians.

Bottom: Loyal troops man an MG 08 machine gun post at the Obelisk in the Karolinenplatz. Munich, May 1919.

Rote Garde
Hoch
d. soz. kommunistische
Räte Republik
Luftschiffer Abtlg.
Revolution in München. Waffenausgabe an die Arbeiter-Wehr
am Marsfeld.

Left: A Jäger Leutnant from the Freikorps Bozendorfer, which was formed by Oberleutnant Ritter von Bozendorfer in April 1919. They later joined with Freikorps von Epp and finally were part of Reichswehr Brigade 21. He is dressed in the M 1915 Friedens uniform with the addition of the Freikorps Bozendorfer oak leaves on his collar. He was awarded the Iron Cross First and Second Class and the Wound Badge.

OPPOSITE
Three Freikorpskampfer, Winter 1919-1920. The man on the right wears the hunting horn and number of Saxon Jäger Bataillon Nr. 12 on the epaulets of his great coat, with the insignia of prominent Freikorps leader Major Bischoff's Eiserne Division on his visor cap. The large seated man also wears the Eiserne Division cap badge on his Kratzchen and a tri-color chevron on his great coat which bears no shoulder boards. The man on the left is dressed in an M 1915 Bluse with collar Litzen and is armed with a small revolver in its holster. His Kratzchen has only one cocard, but no Eiserne Division badge. He is a veteran who was awarded the Wound Badge and also has the tri-color sleeve chevron, but facing in the opposite direction. The colors of these chevrons cannot be positively identified, but in the Baltic campaign, units serving with the Landeswehr wore a yellow, red, and black chevron for identification. Further to that, some of Bischoff's Eiserne Division mutinied in August 1919 and went to fight with General Prince Awaloff's "Russian Army of the West." These men appear to be part of that group.

Above: This photograph shows the rather shabby nature of the dress of the Communist revolutionaries. Naval uniforms of the Volksmarine Division, Army uniforms, mixed Army and civilian clothing, and full civilian dress are all evident. Their weapons are predominately Gew. 98s with a sled mounted MG 08 on the cab of the truck. All cocards have been removed from their caps and a man wearing a civilian hat with a rifle on his shoulder is dressed in an old style Army Litewka jacket. Berlin 1919.

Right: A Hauptmann in the Freikorps von Hülsen had seen previous active service as evidenced by the ribbon and badge of the Iron Cross First and Second Class. He is dressed in his 1915 officers tunic with regimentally marked shoulder boards. Generalleutnant Bernhard von Hülsen formed a large and aggressive Freikorps brigade which bore his name and continued fighting into the campaign in Upper Silesia, May 1921. His troops defeated the Poles at Annaberg, 21 May 1921, and was seen by many as the first German victory since November 1918.

Above: The modified A7V type tank numbered "54" and named "Hedi", Berlin 1919. "Hedi" was created from a chassis found at the Tank Training Detachment in Berlin and fitted with spare body armor plates. Machine guns were fitted at each corner of the vehicle. The Totenkopf insignia of the Kokampf armored units is applied in two variations on "54", one with the skull over the bones as on the side and one with the bones crossed behind the skull as seen on the front.

Right: Another view of "Hedi" and her crew. The crew wore a variety of uniforms from coveralls, as seen on the top man, to M 1912 Lederlitewka on the seated man third from the left, to M 1915 Bluse and an M 1915 officer's tunic. The "Panzer-Kraftwagen-Abteilung" later was under the control of Komand der Kampfwagenabteilung which was shortened to Kokampf. Their barracks, the Kraftfahrer-Kaserne, was in Berlin-Lankwitz. The Kokampf was assigned to Korps Lüttwitz and wore the armband "Kampfwagen".

Above: This Hoffmann photograph shows troops of 1.Württembergischen-Freiwilligen Regiment marching on the Karlsplatz Munich, May 1919.

Below: The spring of 1919 brought hard fighting in the streets of Munich. This 2 May 1919 photo by Hoffmann shows a Daimler-Kraftwagengeschütz armed with a cannon and an MG 08/15 with a 100-round Patronenkasten 16 trommel drum magazine. Of particular interest is the Totenkopf insignia painted on the driver's helmet and on another in the rear. The Totenkopf is in a black circle rather than plainly painted as on the vehicles.

Above: A photograph worthy of detailed study. The vehicle is a Russian Fiat-Radpanzer, first version M 1915. The Kokampf insignia is painted below the gun turret and the name "Lottchen" is on the side of the engine compartment. Perched on the upper body to the right of the door is a steel helmet with a nicely painted Totenkopf on the front and another insignia which cannot be completely seen on the left side, but which looks like a machine gunner's badge. On the next vehicle rests a similar helmet. The NCO on the left is in an M 1915 Bluse with a cavalry-style waistbelt. The collar appears darker, has NCO tressing, and the collar badges for the 3rd Marine Brigade von Loewenfeld, and on his sleeve the cuff title "Kampfwagen" above it a Totenkopf in a black circle and above that the machine gunner's badge which might explain the helmet insignia. His visor cap, as with all the others in the photo, shows only one cocard. The man second from the left wears the 3rd Marine Brigade collar badges, an Army style belt and buckle with an early 1919 style fly-fronted, two pocket tunic. The third man from the left wears the same style tunic and an officers leather field belt, the Iron Cross First Class, Silver Wound Badge and below the Iron Cross what looks to be the silver Kampfwagen-Abzeichen, or tank badge. His trousers have the leather reinforcements as worn by the Sturmtrupp in 1918. The last man is in an M 1915 Bluse, cavalry belt, Marine Brigade and Kokampf insignia, a pair of leather gauntlets and leg-wraps with marching boots. On the front of the armored car a small black flag with Totenkopf can be seen. This wonderful photo is by the courtesy of the Tank Museum Collection, Bovington, England. NR 250/A6.

Below: Oberst Ritter von Epp, nearest to the camera, inspects the Panzerwagen-Abteilung of his Freikorps at Ohrdruff, 1919. The vehicles are Daimler KDs. The black diamond painted on the sides of the armored cars is the same shape as the backing of the sleeve insignia for the Freikorps von Epp snarling lion head. Perhaps they had yet to be added or the poor quality of the photo does not show them. The picture is taken from *Das Buch von deutschen Freikorps-kampfer* by Ernst von Salomon, a 496 page book on the history of the Freikorps as told by many of its members and commanders. The book was published in 1938 and tends to lean towards Nazi ideals.

Above: A company of new volunteers of the Freikorps Lichtschlag line up for inspection. Formed by Hauptmann Lichtschlag in December 1918 in Westphalia, it gained a strength of three infantry battalions, one heavy and one light artillery section. They were involved in the heavy fighting which took place in the Ruhr and Western Germany. One of their units, the Hasenclever Battery, was massacred at Wetter-on-Ruhr in March 1920. This photograph was in all probability taken during the recruiting time in late 1918 or early 1919. Some of the men wear the unit's Freikorps badge, a rampant horse within a shield surrounded by a wreath, on their M 1915 Bluse. Their general appearance does not show a great deal of military sharpness which in all likelihood improved as the unit matured.

Below: An officer and his groom of the Deutsche Schutzdivision (Infanterie Division 31). Both wear the unit Kragenabzeichen which consisted of a sword through an oak-leaf wreath. The officer's tunic bears shoulder straps, the Iron Cross First Class, has four pockets, two breast and two lower slash style, and is fastened by an Imperial officer's leather belt. The enlisted man's apparel is M 1915 issue and his boots are M 1915 Einheits-Kavalleriestiefel.

Right: Hauptmann Rudolf Berthold was born 24 March 1891 in northern Bavaria, served in the Imperial Air Service, scored 44 victories, ended the War with JG II and was awarded the Pour-le-Mèrite in October 1917. On 4 April 1919 he founded the Freikorps Eiserne Schar which was also known as the Freikorps Berthold. The strength of the unit grew to 27 officers and 800 men, making up four infantry and one machine gun companies. Their actions included the conquest of Munich after it was taken by the Communists, the support of the Iron Division's retreat from the Baltic area and also in the Kapp Putsch. Berthold was murdered by the Communists, some say choked by the ribbon of his Pour-le-Mèrite, after which most of his men joined the Loewenfeld Marine Brigade. In this well-known photo, Berthold in an M 1915 Bluse wears his Pour-le-Mèrite, pilot's badge, Iron Cross First Class, and his unit's arm badge of a "B" encircled by a wreath. Also seen is a claw-buckled waistbelt supporting a holstered P. 08 on his left hip.

Above: Polizeihauptmann Phillip, who was Berthold's last adjutant, 1919-1920, is pictured here as a police captain wearing his many awards which include both Iron Crosses, and the next to last on the medal bar, the Medaille der Eiserne Division, with the Totenkopf in full view. His dress uniform is that of a police officer of the late 1920s and early 1930s.

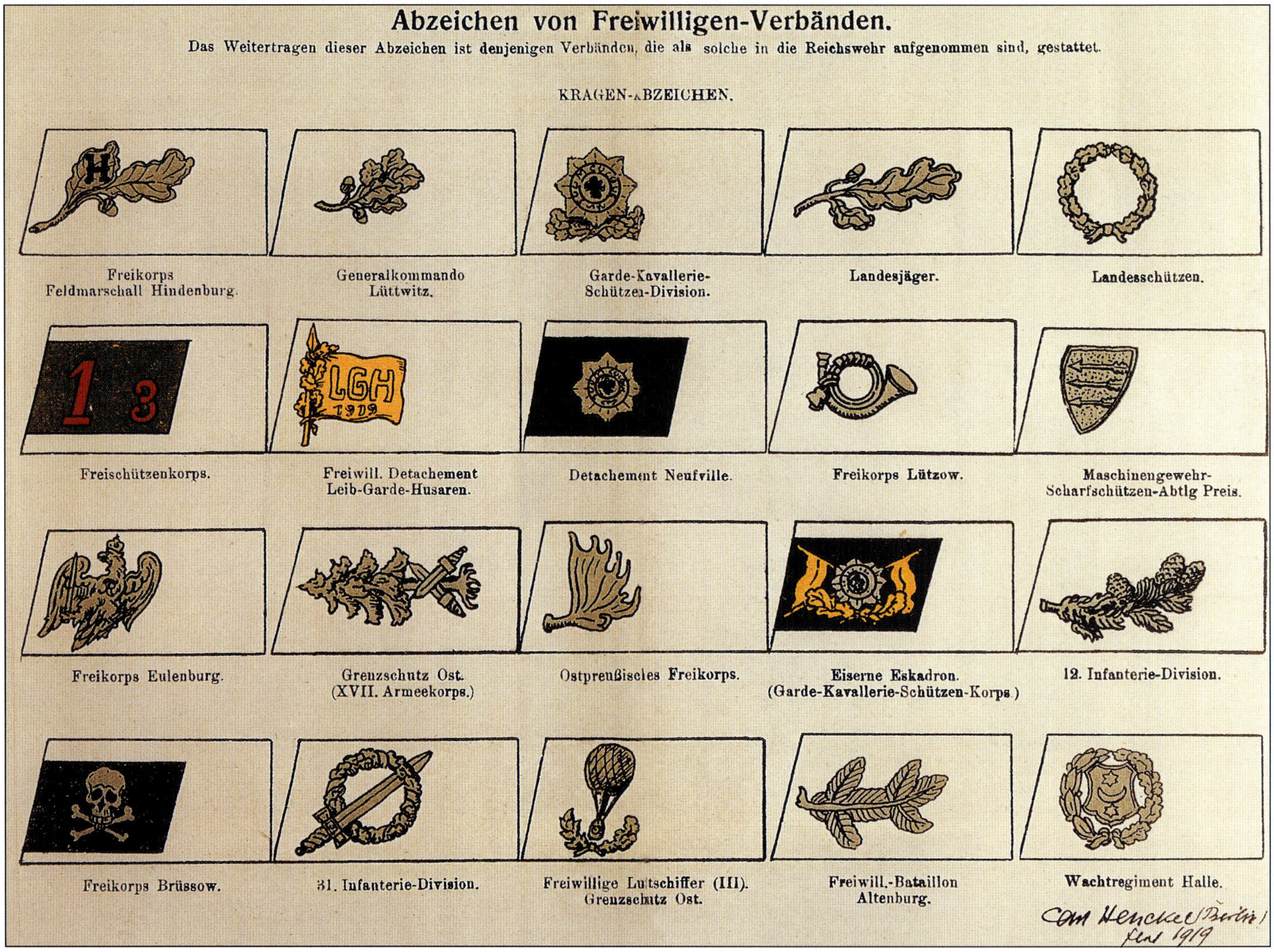

From the second part of the 1919 book entitled *Die neue deutsche Reichswehr, Die Freiwilligen-Verbände und ihre charakteristischen Abzeichen* by the military artist Carl Henckel, come the following three plates showing 64 of the more than one hundred known badges of the volunteers, or Freikorps, of 1919-20.

Abzeichen von Freiwilligen-Verbänden.
KRAGEN-ABZEICHEN, ferner:
Freiwill.-Verbände Anhalt. Inf.-Regt. Nr. 93.
Freiwill. Inf.-Bat. Bremen.
Badisches Sturm-Bat. Ober-Ost.
Freikorps Severin. (Minden.)
Freikorps Dohna.
Westfäl. Jäger-Freikorps.
Freiwillige Grenadiere Nr. 2.
Freiwillige Grenadiere Nr. 9.
Freikorps Schleswig-Holstein.
Hessisch-Thüringisches Freikorps.
Freiwill.-Detachement Tüllmann.
Freikorps Breslau.
Freikorps Düsseldorf.
Grenzschutz-Batterie v. Medem.
Freiwill.-Abteilung Osterroth.
Detachement Grothe.
Freiwill.-Bataillon v. Klüfer.
Grenzschutz-Kommando Suwalki.
Reg. Schutztruppe Bremen.
Freiwill. Hannoversche Jäger.

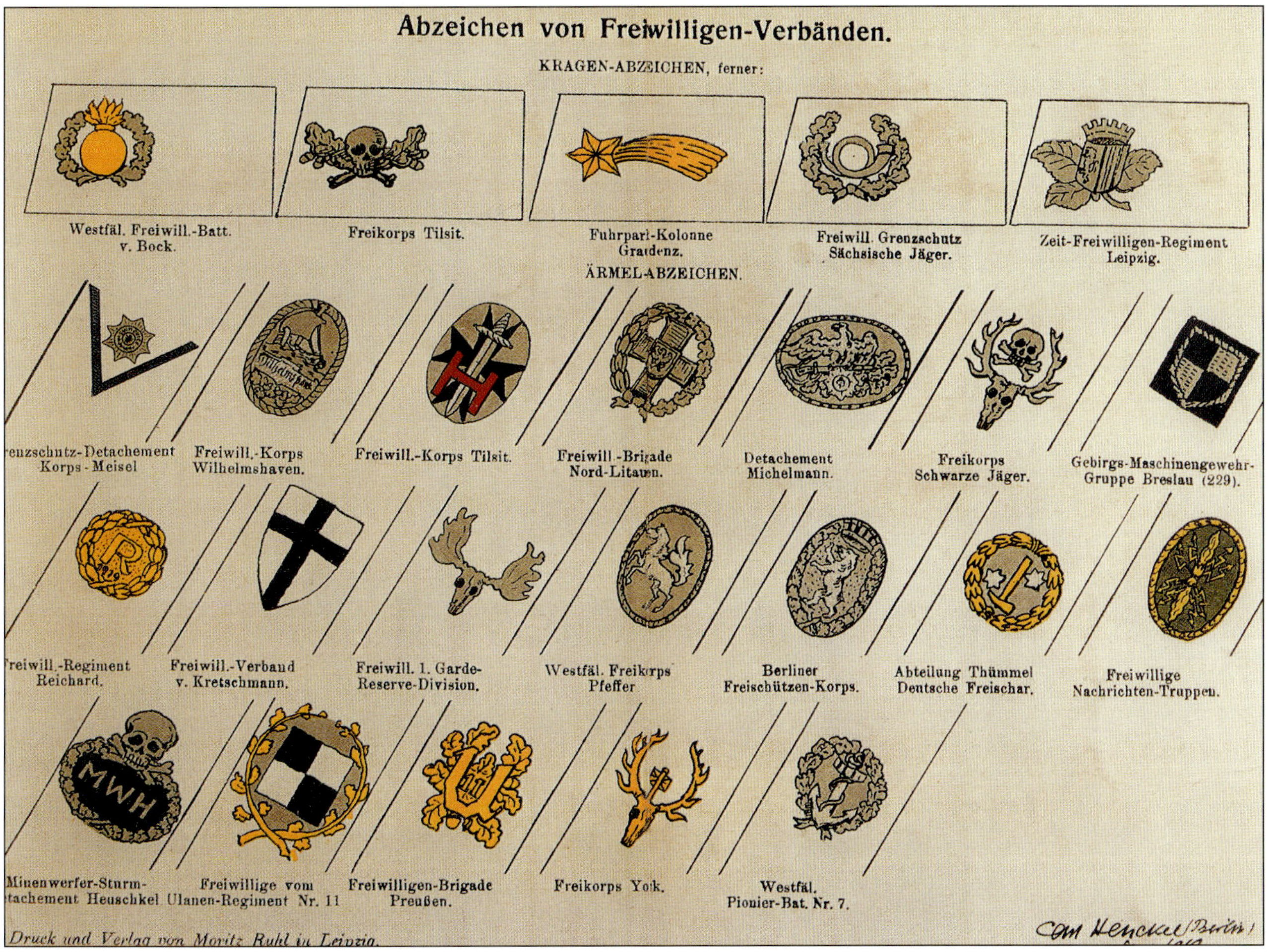
Abzeichen von Freiwilligen-Verbänden.
KRAGEN-ABZEICHEN, ferner:
Westfäl. Freiwill.-Batt. v. Bock.
Freikorps Tilsit.
Fuhrpark-Kolonne Graudenz.
Freiwill. Grenzschutz Sächsische Jäger.
Zeit-Freiwilligen-Regiment Leipzig.
ÄRMEL-ABZEICHEN.
enzschutz-Detachement Korps-Meisel
Freiwill.-Korps Wilhelmshaven.
Freiwill.-Korps Tilsit.
Freiwill.-Brigade Nord-Litauen.
Detachement Michelmann.
Freikorps Schwarze Jäger.
Gebirgs-Maschinengewehr-Gruppe Breslau (229).
reiwill.-Regiment Reichard.
Freiwill.-Verband v. Kretschmann.
Freiwill. 1. Garde-Reserve-Division.
Westfäl. Freikorps Pfeffer
Berliner Freischützen-Korps.
Abteilung Thümmel Deutsche Freischar.
Freiwillige Nachrichten-Truppen.
Minenwerfer-Sturm-tachement Heuschkel
Freiwillige vom Ulanen-Regiment Nr. 11
Freiwilligen-Brigade Preußen.
Freikorps York.
Westfäl. Pionier-Bat. Nr. 7.
MWH
Druck und Verlag von Moritz Ruhl in Leipzig.

Above: Members of Freikorps Schöffelding pose for this 22 April 1919 photograph. This unit was undoubtedly part of another larger unit as the name does not appear in any of the Freikorps lists. One of the most complete of these lists was compiled by Helmut Reuther and was published in several issues of the German publication *Feldgrau* in 1957, entitled *Freikorps und Grenzschutzformation 1918-1924*. Accompanying this list were 17 pages of hand-drawn Abzeichen and flags. These men wear a mixture of M 1915 Bluse and tunics with falling collars and exposed buttons. Five of the seven men carry the numeral "9" on their collars; the left one is applied backwards. The man standing on the left in the photo has his, in what would seem to be the proper order. Armbands of white and yellow are worn by some, as are double cocards on the Mützen of three and singles on the others. Two Ersatz bayonets have been stabbed into the table for a warlike effect.

OPPOSITE

Top: Men of a mixed bag of units, as shown by their insignia, pose in a rural garden setting. The man on the left is a member of the Freiwilligen Grenschutz Ost as indicated by his collar insignia. On his sleeve above his machine gunner's badge is a light stripe designating an Unteroffizer. His uniform is an M 1915 Bluse with dark shoulder straps and he is armed with an M 96 Mauser pistol carried in its shoulder stock holster supported by an Imperial issue waist belt. The second man is equipped with an M 16 helmet, Gew. 98 rifle, M 08 cartridge boxes, an M 1915 Bluse with no collar insignia, but with dark shoulder straps. The man next to him is similarly equipped, but wears an M 1910 tunic and marching boots. The man on the right of the photo is a member of the Baddisches Sturm-Bataillon Ober-Ost as indicated by the oak-leaf supporting a shield worn on his collars. He also wears an M 1915 Bluse and laced boots.

Bottom: Another photo postcard worthy of study. The majority of these men wear the Ulan M 1915 gray Ulanka, many with the insignia of Freikorps von Epp on their sleeves. Above the von Epp insignia the Reichswehr rank insignia is present with the oval Reichswehr unit Armspiegel insignia with the number "21". One man seated on the arm of the bench has been awarded the Iron Cross, both Classes, and retains Imperial-style shoulder straps on his Ulanka. The man kneeling on the far right has what looks like Garde Litzen on his collar, while none of the other Ulans do. The visor caps vary from those with the single cockade of red, white and black and the Reichswehr eagle surrounded by a wreath of oak leaves to the Prussian style of two cocards. The oval Armspiegel "21" would indicate that these men are part of 7. Division, Bavarian Infanterie-Regiment Nr. 21.

Above: A mixture of insignia and uniforms is worn by members of these Reichswehr troops. Uniforms consist of the M 1915 Bluse, the 1919 style tunic with patch pockets and a 1919 pattern with flap pockets. Most wear the Armspiegel for Infantry Regiment Nr. 42 above their rank insignia, as well as a black diamond on their left sleeves which is unidentified. Three have the narrow early 1919 style shoulder straps, some have collar Litzen, so do not, some wear visor caps with Reichswehr insignia, some with the double cocards and one second from the right, an Edelweiss on the side and no cocards. He also has Freikorps collar insignia which are from Freikorps Halle formed in that city 4 March 1919. They later became Sicherheits Polizei Bataillon Halle.

OPPOSITE

Top: A rather poor photograph of the 2. Kompanei of the Jäger Freikorps. The unit, formed 11 March 1919 in Wilhelmshaven, consisted of the Schwarze Brigade and the Schwarze Jäger Brigade. Strength reached 21 officers and 738 NCOs and men consisting of four infantry companies, one machine gun company, and pionier, communications, and mortar units. It later became the I. Reichswehr Schützen Regiment Nr. 18 in Behrenfeld. The insignia varies from man to man, most of whom are wearing the unit arm insignia of a stags head with a Totenkopf between the antlers. Litzen are on the collars of the M 1915 Bluse which also bear shoulder straps with light piping, and several wear the badge of the Garde Kavallerie Schützen Division. The light cap bands on all men have a cavalry flavor, but no solid evidence can confirm that assumption. There is a lineage connection with old marine units, and they finally became part of Reichswehr Infantry Regiment 18.

Bottom: The Freikorps had a number of flying units. This photo shows members of Flieger-Abteilung 418 taken in Glogau, Silesia, February 1919. Among those in the unit with Imperial Air Service backgrounds were Oberleutnant Oscar Freiherr von Boenigk, a Pour le Mèrite winner with 26 victories. He is standing fifth from the left in the front row. Other members of Fl. Abt. 418 in the picture are Ltn. Rudolf Rienau, Jasta 19, 6 victories; Ltn. Herman Becker, Jasta 12, 23 victories; and Ltn. Wilhelm Sensch, Jasta 19, 5 victories. All of these men have retained their Imperial caps and uniforms, continuing to wear their pilot's badges, and in one instance, an observer badge is seen on the man third from the right. An abundance of Iron Cross First Class and Wound Badges can be clearly seen in the picture.

Freikorps
„Schwarze Jäger"
2. Komp.

GARMISCH - PARTENKIRCHENER
VOLKSWEHR bringt gefangene
ROTGARDISTEN"

OPPOSITE

Top: A rather somber group consisting of the staff of Selbstschutz Regiment Schlesien-Rossbach is photographed in 1919. The Silesian campaign produced a large variety of uniform styles as seen in this picture. A Husar Attila in the front row is joined by Imperial and Russian styles, light-colored jackets with shirts and ties worn underneath, and even a civilian staff member appears second from the left. Ltn. Gerhard Rossbach, fourth from the right, one of the Freikorps' most active leaders, became involved with the early Nazi party, but had a falling out with Hitler, later dropping into obscurity.

Bottom: Captured members of the Rote Garde are marched into captivity by members of the Garmish-Partenkirchener-Volkswehr. In the group of captors is an interesting mixture of Bavarians in native costume, probably Freikorps Werdenfels, and uniformed steel-helmeted Freikorpskampfer. The man in the front row nearest the camera in an M 1915 Bluse was awarded the Iron Cross First Class, the Wound Medal, and wears a von Epp sleeve badge. The Rote Garde are dressed in a variety of military and civilian clothing and seem none too happy with their plight.

Above: An unfortunately unclear photo taken in Munich shows members of Freikorps von Epp. In the background are the black, white and red national flag and the white and blue diamond flag of Bavaria. Their helmets have a white band above the brim and a white "E" for Epp above that. The von Epp sleeve badge can be clearly seen on their M 1915 Bluse, M 1908 Patronentasche are braced by bread-bag straps. Gew. 98s and grenades complete their equipment.

From the author's collection, a summer weight tunic of a Leutnant in the Freikorps von Hülsen. The von Hülsen oak-leaves have been applied over the Imperial officer's collar Litzen. The epaulets have white infantry underlay with black Prussian flecking and the pocket buttons are standard gray Imperial crown style. Also of interest is the Freikorps oak-leaf insignia applied to the medal bar, the Iron Cross First Class and the Schlesien Campaign award.

1914
1918

Left: A photo postcard of a typical Reichswehr Schütze in early 1919. This M 1915 Bluse has had the shoulder straps removed and the man is equipped with an M 16 helmet, M 08 pouches, Bz. 15 grenades, Gew. 98 with a 98/05 bayonet, leg-wraps and laced marching boots.

OPPOSITE
A great studio portrait of Willy Westendorp of the 3. Batterie Feld-Artillerie Regiment 80 of the Freikorps-Eulenburg, Frankfurt am Oder, dated 7 June 1919. This large Freikorps was recruited by Major Graf zu Eulenburg and formally established on 15 February 1919 in Potsdam. The unit was comprised of Infanterie, 6 Kompanie, Maschinengehr and Minenwerfer Kompanie, Dragonen, Feld-Artillerie, 2 Kompanie, and a Pionier Kompanie. They fought in the Baltic and in the border disputes. Our subject wears an M 1915 Bluse, mounted troops boots, a visor cap with a black band for artillery and both cocards, an Artillerie Schützenabzeichen 6th class, ribbon for the Iron Cross Second Class, a Wound Badge, and the Freikorps collar insignia which is an eagle holding a sword with a script "R" on its breast.

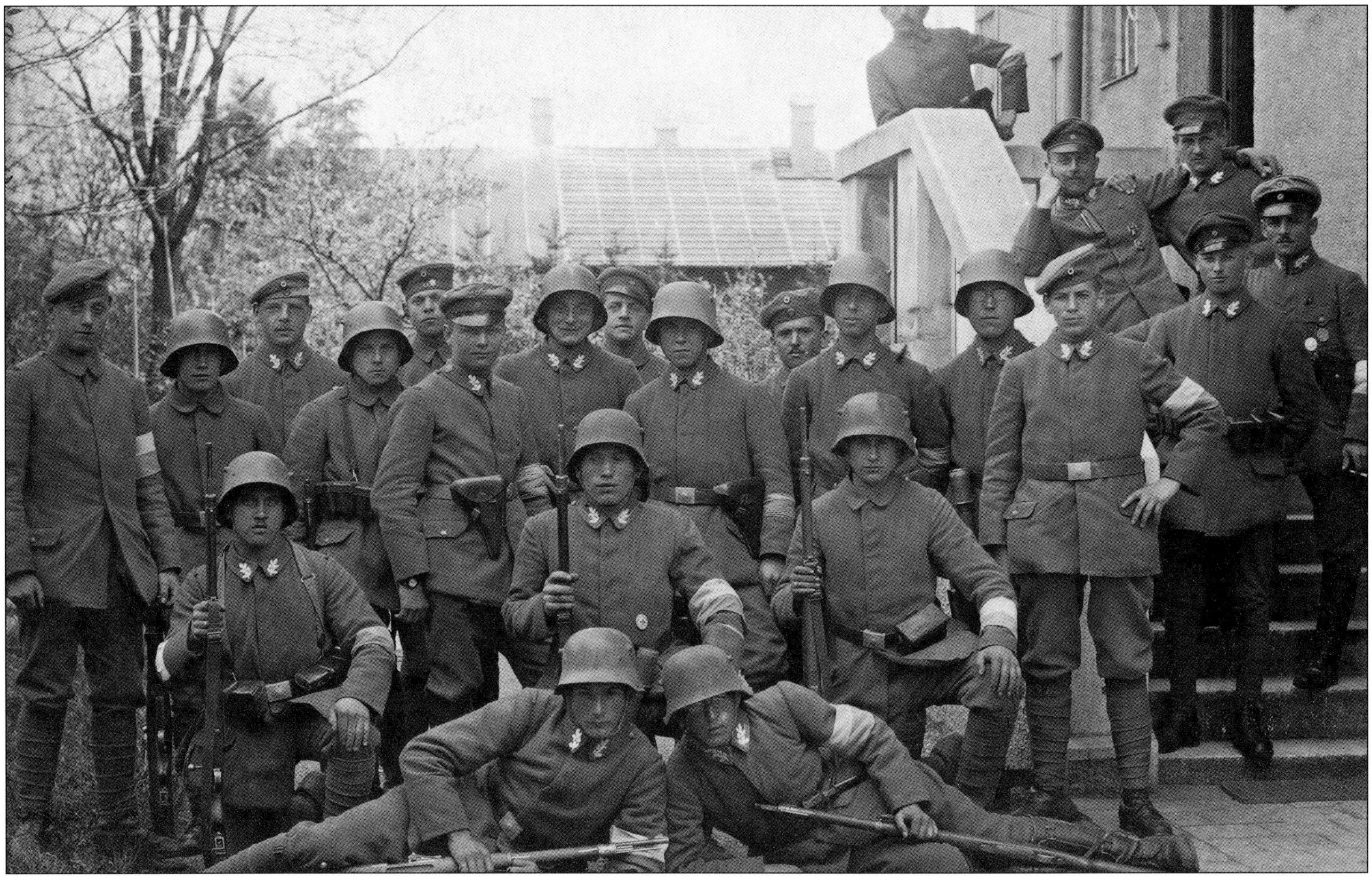

Above: A group photo postcard of members of the Oberschlesisches Landjägerkorps. The unit's distinctive oak-leaf collar badges are worn by most and a gold or yellow and white armband is also evident. All but one man on the far right wear the M 1915 Bluse, with black and silver Wound Badges, Iron Crosses, and NCO collar discs are very much in evidence. Holstered P. 08s, Gew. 98s, Kar. 98a's, Bz. 15 grenades and an unidentified revolver on the man far right make up the weaponry of this unit. The prone man on the right is armed with Kar. 98a and a trench knife and wears Ledergamaschen rather than the cloth leg-wraps worn by the others.

OPPOSITE

Top: Freikorps Werdenfels were the most colorful of the Freikorps units, dressing entirely in Bavarian costume. They are shown in this Hoffmann photo, taken in Munich May 1919, drinking beer and enjoying hot food from the "Gulaschkanone". Close inspection shows the word "Werdenfels" embroidered on the suspenders of the lederhosen. The unit was formed by Major Ritter von Reiss on 21 April 1919 at Garmish-Partenkirchen. Their strength was two companies of about 250 men and most of their fighting took place in the liberation of Munich. Their insignia was a white and blue armband with a von Epp diamond and lion device in the center and the printed words "Freikorps Werdenfels" on the white portion of the armband.

Bottom: The Freikorps Werdenfels doff their hats in salute to the photographer in this May 1919 picture taken in Munich. The Werdenfels flag is on the left. A guitar player in the center and a general gaiety seem to overwhelm the somber presence of Gew. 98s and an MG 08 on its sled mount.

FREIKORPS "WERDENFELS"
ZUR BEFREIUNG MÜNCHENS"

Above: The Werdenfels forming up in Munich, May 1919.

Below: The Werdenfels marching through the streets of Munich, May 1919.

As the Communist threat grew, the need for additional volunteers for the armed forces grew also. The Freikorps Deutsche Schütz-Division under the command of Generalmajor von du Lippe was recruiting volunteers for the 31. Infanterie Division and produced this and the following recruiting poster.

Recruiting poster for the Freikorps Deutsche-Schütz-Division, the 31. Infanterie Division. The collar insignia for this division was a sword piercing an oak-leaf wreath as seen in an earlier illustration.

A recruiting poster for the Garde Kavallerie Schützen Division offering opportunities to officers, NCOs, and men of all branches or specialties. The association with the Landsknecht image is quite clear in this poster.

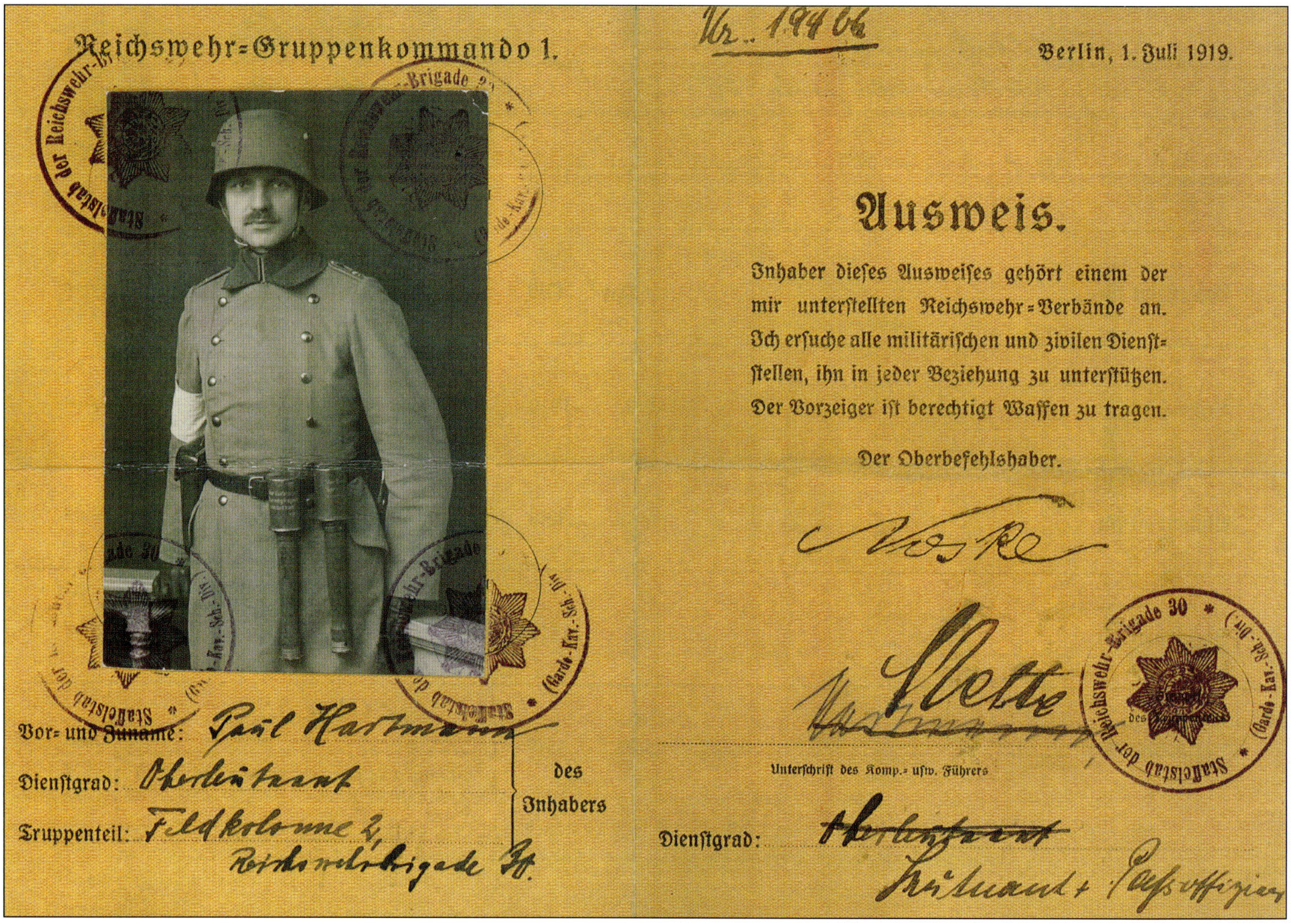

Reichswehr-Gruppenkommando 1.

Berlin, 1. Juli 1919.

Ausweis.

Inhaber dieses Ausweises gehört einem der mir unterstellten Reichswehr-Verbände an. Ich ersuche alle militärischen und zivilen Dienststellen, ihn in jeder Beziehung zu unterstützen. Der Vorzeiger ist berechtigt Waffen zu tragen.

Der Oberbefehlshaber.

Noske

Vor- und Zuname: Paul Hartmann

Dienstgrad: Oberleutnant

Truppenteil: Feldkolonne 2, Reichswehr-Brigade 30.

des Inhabers

Unterschrift des Komp.- usw. Führers

Dienstgrad:

Above: The Ausweis of Oberleutnant Paul Hartmann Garde Kavallerie Schützen Division, Reichswehr-Brigade 30 who was a member of Feldkolonne 2. of the Reichs-Brigade 30. In this formal photograph he poses with hand grenades, P. 08 and steel helmet. The white armband is unidentified. The Garde-Schützen-Division, created at the end of 1918, was commanded by Generalleutnant von Hoffmann. Among the old army units which made up the Division were Garde Kürassiere Regt., Leib Kürassiere Regt., Dragon Regiment 8, Husar Regiments 8 and 11, Ulanen-Regiment 5, Jäger Regiments zu Pferde 2 and 6, and part of 4. The Garde M.G. Abteilung, 3. Garde F.A. Regiment, Pionier Bataillon 412 and Minenwerfer-Kompanie 385. They fought in Berlin December 1918, January 1919, March 1919, and April-May 1919 in Munich.

OPPOSITE

Top: One of the many Freikorps which rallied to the German Republic to assist in the suppression and prevention of the Communist takeover was Tank-Abteilung Körting. Formed by Leutnant Georg Körting who was to somehow manage to collect varied armored cars of different nationalities and no less than five tanks. There were three captured British Female MK IVs, a captured Whippet and a very modified A7V which was named "Hedi". This unit was later named Kommand der Kampfwagenabteilungen, or "Kokampf", and was stationed in a Kaserne in Berlin-Lankwitz. The Totenkopf insignia adorned their vehicles which took part in the streetfighting in Berlin, Munich, and later in Leipzig. Mechanical troubles plagued the units and it was difficult to have more than half their vehicles in service at one time. This photograph from the collection of the Tank Museum, Bovington, England, shows two detachments, Nr. 101 and 103 of the Kokampf at Berlin-Lankwitz. One detachment on the left is equipped with Ehrhardt E-V/4s and a Benz touring car, while the other unit has Russian Austin's and an unidentified make touring car.

Bottom: This and the following four photographs are of the same Ehrhardt E-V/4, 4 cylinder, 85 H.P. water-cooled armored car as it appeared in 1919. This vehicle is equipped with at least four machine guns, a starting crank, hard rubber tires and is well adorned with the Kokampf Totenkopf insignia.

Above: On the rear of this same Ehrhardt E-V/4, a Pionierei shovel is attached above the towing rings flanking the Totenkopf. Close inspection of the Ehrhardt shows the right rear wheel is different from the right front, and looking under the engine compartment, the left front is different, having no spoke cover. Another view of the same car shows the left rear wheel to be the same as the right rear. **Below**: A left-hand view of the Ehrhardt.

Above: Government troops stand watch over the Ehrhardt, while a curious civilian crowd looks on. **Below**: The location on the street of this car in relation to the photo shown above, shows the car reversing its direction. The wheel marks in the road show this was occurring.

Above: The May 1919 liberation of Munich from Communist control saw many units in action. This tough looking group comes from Detachment Glasser, 5. Komp. I Zug. 19 Infanterie Regiment. The Freikorps oak-leaf motif is shown on the identification slate, but only the white and yellow armbands worn by some to show Freikorps affiliation. The tall officer in the center is in all probability Major Glasser, the unit's commander.

Left: A Sergeant from Bavarian Feld-Artillerie Regiment Nr. 15 poses wearing a Bavarian M 1916 Bluse with a modified collar badge of Freikorps Weckbecker, part of Freikorps Landsberg, whose commander, Major Mack, formed the unit 15 April 1919. They later became part of Reichswehr Infanterie Regiment. 42, III. Bataillon. The Weckbecker insignia has been modified by the removal of the letter "W" between the leaves.

OPPOSITE

Top: Freikorps Detachment Probstmayr was formed by Major Probstmayr, 5 April 1919 in Donauwörth. The unit consisted of about 500 men and their insignia was a white and blue armband which can be seen in this photo postcard. Weapons seen here are Kar. 98a's and Bz. 15 grenades. The man seated in the center has a Karpathenkorps device on his visor cap, an echo of his prior service.

Bottom: More men from Detachment Glasser gather around the "Gulaschekanon" for a hot meal and a mug of beer by two of them. The large man in the center is dressed in an M 1910 tunic with "19" on the shoulder straps. Others are in the M 1915 Bluse, while sweaters and shirtsleeves are prominent on the three men on the left. Two men wear Freikorps oak-leaves on their collars, many have white and blue armbands and all have removed the national cockade from their visor caps and Mützen. The location is in front of the Hofden Implerschule in Sendling, mid-May 1919.

München
1. u. 2. Mai 1919.
Kustermann

Above: Oberstleutnant Faupel on horseback reviewing his Freikorps Faupel, also know as Freikorps Görlitz, or Schlesische Jäger-Brigade, which he formed in January 1919. The unit grew to considerable size with a strength of 94 officers and 2300 men. There were four infantry batallions, four machine gun companies, one Minenwerfer company, one Radfahrer, or bicycle company, one squadron of cavalry, and four artillery batteries. They fought in Görlitz, Magdeburg, Dresden, Munich, and the Ruhr fighting of 1920. They became Reichswehr Grenadier Regt. Nr. 10. The unit wore a plain double chevron or a chevron in a wreath on their lower sleeve for their distinctive insignia. In this review, Faupel was joined by Generalleutnant von Friedburg (in the light Mantel) who commanded the 2. Gardedivision. The Feldherrnhalle can be plainly seen in the background.

OPPOSITE

Top: A Field Artillery battery passed in a review by Oberstleutant Faupel and Generalleutnant von Friedburg who are in the Odeonplatz, Munich, 6 May 1919.

Bottom: A clear view of Genlnt. von Friedburg reviewing a Freikorps Faupel Maschinengewehr company. At least four men in the rear of the formation are wearing MG 08 dragging straps as a badge of their specialty.

Left: The 1. Garde-Reserve-Division adopted a elk skull and antlers for its insignia as shown on the sleeve of this NCO. The 1. Garde-Reserve-Division consisted of units from 1. Garde-Reserve-Brigade, Garde-Reserve-Regiment 64, Detachment von Brandis (Infanterie Regiment 24), Bataillon von Grafenstein, Bataillon von Maltzahn, Staff of Dragon Regiment 2 (Major Graf Yorck), 1. Garde Regiment Feld-Artillerie Detachment Stever and the II. Garde Fuss-artillerie Regiment. They were commanded by Generalmajor Tiede and saw most of their action in the Baltic. Our subject was awarded the Iron Cross, both classes, and the black Wound Badge. A cavalry-style belt supports a fighting knife, the scabbard just visible on the right. The uniform is an M 1915 Bluse with collar Litzen. The chevron below the unit insignia is not identified, but may well pertain to one of the sub-units, as well as the devices on his shoulder straps.

OPPOSITE

Top: The reason for the gathering of these Bavarians is obscure, but it portrays an interesting mixture of Freikorpskampfers and civilians, young and old. The dapper gentleman on the far left wears a soft crowned visor cap with a single Bavarian cockade, a short lapelled coat over his Bluse with the Freikorps oak-leaf design collar Abzeichen, leg-wraps, laced marching boots, and a cane. A Bavarian M 1916 Bluse, plain Bluse with upper slash pockets and M 1915 Bluse are all evident. The two musicians lend a festive air to the group.

A Freikorps private poses for a formal photo wearing an M 1915 style Bluse with the same collar insignia worn by members of the Freiwilligen-Landsjäger Korps, Detachment Probstmayer, Freiwilligen Bogendörfer, or the 17. Infanterie Division.

Above: Early in the Communist takeover of Munich, Red army revolutionaries occupied many of the city's key buildings. Here we see the Reds posing before the Luitpoldgymnasium. Close study of this photo shows the faces of old and young wearing a mixture of uniforms, Bavarian M 1916 Bluse with Bavarian blue and white tressing on the collars, M 1915 Prussian Bluse, tan jackets with button fronts worn over civilian and military clothing, plain civilian dress with military visor caps, as well as many red armbands. One man holds a drum while others are armed with Gew. 98s, Bz. 15 grenades, and the ever-present MG 08 on its sled mount. This photo and the one to follow were later marked by the police in their effort to track down Communist agitators after the revolution had been quelled.

OPPOSITE

Top: Another tough looking group of revolutionaries and a serious young man are pictured in what seems to be a barracks room. Weapons hang on the wall and bedding is visible on the far left of the photo. The tan wind jackets seen in the previous picture are here again on five of the men and when studied carefully, look to be mountain jackets as issued to Bavarian Gebirgsjäger troops. It would seem that they were "liberated" from some government stores and passed out to the Red troops. M 1915 Bluse, red armbands, Mützen and visor caps with a single cocard, plus one civilian cap, make up the balance of the uniforms and accessories. The police were able to identify the following pictured here: 1. Michael Seidl; 2. Josef Theiler; 6. Ludwig Seehuber; 7. Franz Grabmeier; and 9. Bartl Steiner.

Bottom: Civilians are being questioned by members of 1. Württembergischen-Freiwilligen Regiment in Munich, late April 1919. This photograph contains a wealth of information as to the variety of weapons and equipment used by the Freikorps. The following can be seen: M 1918 and M 1916 Stahlhelm, Gew. 98s, Kar. 98a, and on the man on the right, Patronentasche 95 cartridge box, Beilpicke 98 axe, a Spaten 98 entrenching tool, and a Feldflasche 1915 canteen, an M 1915 Bluse with a white and blue armband, leg-wraps and laced marching boots. The man in the Feldmütze shows the Freikorps oak-leaves on his M 1915 Bluse, as well as a white and blue armband. He is armed with a trench fighting knife, an Ersatz bayonet and a Kar. 98a. The other two men ward off the cold spring weather in Feldgrauer Mantels M 1915.

Above: Murder, looting and general mayhem took place daily on the streets of Munich during Revolutionary takeover and subsequent freeing of the city. Government and Freikorps troops were called upon not only to fight the Communist scourge, but also assist the police and apprehend wrongdoers. Here we see the capture of accused murderer Johann Lehner, iron lathe operator, by troops of the 1. Württembergischen Freikorps, 3 May 1919. The Württembergers are variously uniformed, but those with weapons, Kar. 98a's, are equipped with the special Württemberg war-time Patronentaschen which were enlarged and partitioned to take extra ammunition for combat use. The fate of Herr Lehner is unknown.

OPPOSITE
Top: The Rote Garde mounts up to be transported to a position to engage the Freikorps. This photo clearly shows the civilian soft cloth visored cap which was the cap of choice worn by the Communists. The type of costume seen here was widely worn in the heavy fighting in the Ruhr. Weapons are Gew. 98s and a two-man crew with an MG 08/15. The ammunition carrier wears a red armband and is holding two Patronenkasten 15, each holding 250 rounds, while the gunner holds the gun itself. The double-barrel case for spare barrels is slung over his shoulder.

Bottom: Members of Freikorps Feldmarschall von Hindenburg surround their namesake. The commanding officer and founder was Hauptman Otto who formed the unit in Kolberg, 8 March 1919. They fought mainly on the eastern borders and had a strength of 25 officers, 167 NCOs and 573 men. The troops were made up of one Sturmbattailon, 3 Infantry companies, one machine gun company, one Minenwerfer company, a squadron of cavalry, a communications section and a recruiting station. They later formed parts of Reichswehr Infantry Regiment 110 and Reichswehr Cavalry Regiment 10. Their insignia consisted of oak-leaf collar devices with an applied "H' and a sleeve insignia of an "H" within a wreath.

Above: Commanding Officer Major Bischoff, front row center, and the staff of the "Eisernen Division" in the Baltic, 1919. The Eisernen Division was one of the toughest of the Freikorps, consisting of the following units: 1st, 2nd and 3rd Kurland Regiments, a Jäger Bataillon, a machine gun sharpshooter Abteilung, a cavalry regiment under Major Graf Kanitz, artillery batteries under Major Sixt von Arnim, Flieger Abteilung 427, later addition of F.A. 429, a Minenwerfer unit, two pionier companies, two armored units and Sturmabteilung Rossbach. The majority of the fighting by the Eisernen Division was in the Baltic.

Above: The "Eisernen Division" under review somewhere in the Baltic. The troops on the right are either from the Jäger Bataillon or the Maschinengewehr Sharfschützen Abteilung as indicated by their shakos. They are being reviewed by a highly decorated General wearing a pickelhaube, two Jäger or Maschinengewehr officers, General Fürst Awaloff-Bermondt in the Kossak uniform, the commander of the Freiwilligen Westarmee (White Russian), and next to him, one of his officer's in Russian uniform.

OPPOSITE

Top: In the previous photograph we saw General Fürst Awaloff-Bermondt, commander of the Freiwilligen Westarmee, and in this rather blurry picture taken in late Fall 1919, we see a group of his troops interestingly wearing German uniforms, many with White Russian cap devices, as well as Russian-style shoulder boards. Kar. 98a's and Gew. 98s are both in use, as are holstered pistols and Bz. 1915 grenades.

Bottom: Pictured in 1921 is the flag of Rossbach's Selbstschütz-Regiment Schlesien receiving its battle streamers in 1921. This was part of the larger Freikorps (Sturmabteilung) Rossbach which consisted of thirty-two officers and 960 NCOs and men. In August 1919 their total armament consisted of three field cannon, two light field howitzers, twenty-six heavy machine guns, twenty-eight light machine guns, four light Minenwerfer, three medium Minenwerfer, six flamethrowers, 951 Gew. 98s and Kar. 98a's, ninety-two machine pistols and one armored car. The machine pistols were MP 18s so much admired by the Storm Troops. This is one of the few units for which we have any breakdown of weapons. The unit sleeve insignia was a deer skull surmounted by a cross and set into a red, white, and black chevron.

OPPOSITE

Top: This intriguing picture probably was taken in the Baltic region of Freiwilligen-Westarmee troops. This group wear a mixture of German and Russian-style uniforms and a variety of sleeve badges. The man kneeling on the left is seen in an M 1916 Bavarian Bluse with the ribbon of the Iron Cross Second Class, and on his sleeve; the top badge looks like that of Freikorps von Diebitsch, 948 officers and men, who fought in Litauen. Next, the double-armed cross is that worn by Freiwilligen Westarmee. A drawing in "Feldgrau" shows another Westarmee badge with a third short arm on an angle below the other two seen here. Lothar Hartung, in his *Deutsche Freikorps 1918-1921* shows a photo of this double-armed cross within a circle, attributed to the Westarmee. Below the cross is the badge of a machine gunner sharpshooter which finishes the sleeve ornamentation. The next two men have a definite Russian flavor in their uniform choices, although their caps bear German cocards. The remaining men wear M 1916 Bavarian Bluse, or M 1915 Prussian styles. The man kneeling far right wears the collar button of an NCO, a ribbon for the Iron Cross Second Class, a Wound Badge and the Freikorps von Diebitsch sleeve badge. A rare and most unusual photograph dated *1 Mai 1919. Swvalki.*

Bottom: Freikorps equipment came in many varieties as this picture well illustrates. Bavarian Freikorps Bayreuth men wearing French Adrian-style helmets pose by the Army Museum in Munich during the fighting in 1919. Close inspection shows that there were no insignia affixed to their helmets which one source states came from the Army Museum itself. It seems rather strange that the Museum would have had so many of these helmets in their holdings. The men wear a variety of M 1916 Bavarian Bluse, M 1910 tunics, and M 1915 Mantels. Holstered P. 08s , Gew. 98s, Bz. 15 grenades and large Württemburg-style, as well as M 1909 Patronentaschen, make up the equipment of this group. Many wear white and blue armbands and the mounted NCO carries an M 1890 Stahlrohr-Lanze with the plain white and blue Bavarian pennants as carried by troopers.

Above: Cavalry purportedly from Freikorps Lichtschlag. Ernst von Salomon in his book *Das Buch von Deutschen Freikorps-kampfer* captions this photo "A squadron of the Freikorps Lichtschlag" with the photo source being the Heeresarchiv. Hauptmann Lichtschlag formed the unit 14 December 1918 in Westphalia. The unit consisted of 3 infantry bataillons, one heavy artillery Abteilung, one light artillery Abteilung, and one Kraftwagonkölonne. After September 1919 they were incorporated into Reichswehr Schützen Regiment 107. This cavalry squadron is not included in the table of units mentioned by Helmut Reuther in his history *Freikorps and Grenzschutzformationen 1918-1924* appearing in several parts in *Feldgrau* in the late 1950s. Nonetheless, this is a most interesting picture including on the far left, a man wearing a steel helmet, a fleece-lined coat of Russian origin, and cavalry boots. He is armed with a P. 08 in his right hand with its holster on his belt. Over his shoulder is a Kar. 88 and two Bz. 15 grenades through his belt. The man on the right appears in an M 18 cutout helmet, M 1915 Mantel with shoulder straps and a Freikorps sleeve badge which for the Lichtschlag troops was a shield containing a rampant horse surrounded by a wreath. A Kar. 98a is slung over his shoulder. Others in the squadron also wear M 18 cutout helmets and carry M 1890 Stahlrohr-Lanze with black and white Prussian pennants.

Above: Reserve Officer candidate Johannes Fichler pictured while serving with Freiwillige Detachment von Randow. This unit was formed by Hauptmann von Randow 5 January 1919 in Schaulen. It was divided into two groups, Südgrüppe and Nordgrüppe and was involved in the fighting in the Baltic. The south group included a battery of Kannone, Infantry, an MG Abteilung, railway troops, and cavalry. The north group had infantry, two machine gun companies, a cavalry squadron, Feldhaubiten (howitzer) and 10.5 cm Feld-Kanone batteries, a Pionier company and a communications unit, for a total force of 2653 men. Fichler served with the Imperial Marine Artillery in Flanders during the war and was awarded the Bavarian Military Merit Cross Second Class, with swords, the Iron Cross Second Class, and the Oldenburg Friedrich August Cross Second Class. For his Freikorps service, he won Detachment von Randow Steckkreuz First Klasse and the Baltic Cross which are worn on the left breast of his Naval feldgrau Bluse. On his sleeve is the Eastern Frontier Defense Forces shield of the Order of the Teutonic Knights. On his Naval officer's cap is the Totenkopf badge of the Eiserne Division.

OPPOSITE

Top: Officers of III Marine-Brigade von Loewenfeld. Freg. Kpt. von Loewenfeld formed this unit from Naval forces in Kiel and consisted of Marine Regiments 5 and 6 which included machine gun, Minenwerfer and infantry troops, a Sturm bataillon, a sharpshooter company, armored car company, pioniers, field and mountain artillery, communications and a Flieger Abteilung rounding out their 6000 man force. They fought in Berlin, Oberschlesien, and the Ruhr, and were disbanded on 16 June 1920. The insignia of this unit was a fouled anchor surrounded by a ribboned wreath of oak leaves. Those pictured are L/R: Kptlt. Classen, commander of II Batt. Marine Regiment 6; Kptlt Carls, commander of I Batt. Marine Regiment 6; Major Bruns, commander of Marine Regiment 6; Major Kloebe, commander of Marine Regiment 5; Korvettenkapitän Kolbe, commander of I Batt. Marine Regiment 5; Kptlt. von Arnould de la Perriere, commander of the Sturmbataillons; and Kptlt. von Fischel, commander of the II Batt. Marine Regiment 5. Major Kloebe and Kptlt. von Arnould de la Perriere are wearing their Pour le Mèrite awards. Former U-boat officers von Fischel and de la Perriere wear their U-boat badges for Imperial service.

OPPOSITE

Bottom left: This formal photo postcard is simply signed "Friedrich, February 1919" and depicts a Reichswehr Marinekorps Matrose, or sailor, in a Feldgrau M 1915 Bluse, feldgrau trousers with leg-wraps, and a feldgrau Mütze with a band which reads "Reichswehr Marinekorps".

Bottom right: A member of the Eiserne Division, as indicated by the Totenkopf on his visor cap. The visor cap also carries a cocard which does not appear to be a normal style. Under magnification it cannot be seen clearly enough for identification. His Mantel, or overcoat, is an M 1915 with Imperial crown buttons.

Left: NCO Joseph Willmeroth poses for a formal portrait in an M 1915 Bluse with shoulder straps having a triple band around them, a ribbon for the Iron Cross Second Class, a black, white and red shield device on his sleeve which signifies service in Freikorps Hasse, as does the Brunswick-style Totenkopf on his cap. Freikorps Hasse was formed by Oberst Hasse and consisted of Freiwillige Regiment Hasse, Minenwerfer-Kompanie Hasse, one artillery Abteilung, and one heavy artillery battery. The unit strength grew to 114 officers, 355 NCOs and 1549 men. They fought in Marberg, Berlin, and the borders of Oberschlesien, and in February 1920, they were absorbed into Reichswehr-Infanterie-Regiment 108.

OPPOSITE
Two members of Freikorps Oberland on guard duty equipped with M 1915 Bluse, M 1909 ammunition pouches and Gew. 98s. The Edelweiss collar insignia is that of Freikorps Oberland which was formed by Major Ritter von Beckh on 12 April 1919 and consisted of infantry, machine gun, pionier and recruiting troops numbering 120 officers and 1530 NCOs and enlisted men. They fought in Munich and as a Self-Protection troop (Selbstschütznerband) in Oberschlesien in 1921. Later it became III. Reichswehr Schützen Regiment 42.

This interesting photo was taken in July 1919 and poses some questions as to which unit this man belonged. The collar attachment looks to be the same color as the uniform and has a large chainstitched "2" and a smaller "10", while under magnification the sleeve badge has the look of that of the Freiwillige Kompanie Osterreicher. The numbered collar attachment is identified in *Archiv für Waffen-und Uniformkunde* as being worn by the "Volkswehr", with no further information. The tunic is an early 1919 "Regulation" pattern, with buttons that are plain pebble-grained. The cap has a gray visor dark band and two cocards.

Left: The collar devices of this freshly equipped man show him to be a member of the Deutsche Schützdivision which was the volunteer part of the 31. Infanterie Division formed by Generalmajor von du Lippe. They fought in Berlin during the March riots and were later made part of Reichswehr Brigade 25. His M 1915 Bluse looks like recent issue, as does the rest of his equipment. The helmet is an M 16 Austrian as shown by the fabric chinstrap and the weapon is a Kar. 98a with an M 98/05 bayonet.

OPPOSITE

Top: Officers, NCOs and men of the Ehrhardt Brigade pose for a group photo all wearing the Brigade's sleeve badge depicting a Viking ship, plus there are other collar badges, rank insignia, and Iron Crosses First Class. The officer in the center wearing breeches with light colored leather reinforcing, wears what looks to be crossed cannon over a single Chevron, the significance of which is unknown. Next to him on the left is the "Speiss", or First Sergeant, with his notebook tucked into the front of his Bluse which also bears the Iron Cross First Class and the ribbon for the Iron Cross Second Class. The specific unit within the Brigade cannot be determined.

In the early days of the Revolution, the Communists wore a variety of uniforms, Naval blue uniforms, Naval feldgrau uniforms, Army uniforms as were worn at the front, without any unit identification, mixed uniforms of military and civilian make-up and plain civilian clothes. On all these uniforms and combinations, a red armband was often seen, as well as small red flags waved as a sign of their intent. The following photographs show the look of the rather scruffy Red forces.

Communist forces enter Berlin, 9 November 1918, two days before the Armistice. Army and Naval personnel can be seen on the rather tired looking truck. MG 08/15s rest menacingly on the cab, a Red flag is being waved and near the rear of the truck, one man waves a pickelhaube by its spike. Many of those in the crowd are amused by the entire affair.

Above: On 5 January 1919, revolutionaries march down Berlin's Lindenstrasse led by a truck carrying an MG 08 on the roof and a mixed dressed group of Red soldiers. Of interest is a marcher with a Gew. 98 over his shoulder wearing a soft fedora hat and a somewhat passé Litewka and dark trousers.

OPPOSITE

Top: November 1918: Revolutionaries surround a Russian 1915 Pierce Arrow armored car, one of two in German hands, in the courtyard of the Royal Palace. The grim face of the Revolution is never more clear than what appears in this widely distributed photo postcard. All cocards have been removed from their headgear, although the Naval cap tallies remain, one matrose wields a Naval cutlass, while others are armed with Gew. 98s, Kar. 98a's, and Bz. 15 grenades. The uniforms worn are a mix of Naval and Army, with a civilian outfit thrown in for good measure.

Bottom: Followers of the Red flag pose by a cannon in the early part of 1919. Here we clearly see the varied dress of the Communist uprisers from full Naval attire to Army uniforms to civilian hats, vests, coats and ties. The two weapons that appear in this photo are a Gew. 88 Commission rifle and a Gew. 71/84.

Revolutionstage in Berlin.
Panzerautomobilbesatzung im Hofe des Schlosses.
Ø 8.
S. u. G. S. i. B.
Phot. A. Groß.

Verteidigung der Roten Fahne

12

IA 6305

OPPOSITE
Top: Revolutionaries occupy the square by the Berlin Stadtschloss. An MG 08 on its sled-mount surrounded by 500 round ammunition boxes and operated by helmeted gunner and served by a Matrose identified as Franz Beirsdorf. In the background stands a man wearing a Mütze without cocards and an M 1910 tunic, armed with a Gew. 98, an ammunition bandolier, and a hand grenade. The date was November 1918.

Bottom: An overheated truck carries armed Communist workers and sailors down the avenue by the Marstall. Berlin, 6 January 1919.

Above: Revolutionary forces occupy the Marstall. Berlin, 9 November 1918.

Below: This group of sailors form a "Matrosenpatrouille" on Berlin's Friedrichstrasse on 10 November 1918. They carry their Gew. 98s slung muzzle down, their caps have the cocards removed and they wear full Naval uniforms and light-colored armbands.

A guard post of the Volksmarine Division stands by the badly damaged Imperial Palace gates, the Schlossportal. The date was 24 December 1918.

Above: Communists march down a wide Berlin street in January 1919. Here again we see the mixed clothing worn by the Revolutionaries. This group is not heavily armed, with only two Gew. 98s in view. Note the white armband on the sailor, who also appears in an earlier picture.

Below: A photo postcard taken in Berlin the first day of the Revolution on Unter den Linden at the corner of Friedrichstrasse, December 1918. This group is universally heavily armed in contrast to those in the preceding photo.

Above: Troops of the old Husar Rgt. 11, part of the Freikorps Garde-Kavallerie-Schützen-Division, pass in review, Berlin 1919. The Division was founded by Generalleutnant von Hoffmann in December 1918, and consisted of cavalry troopers from the Garde Kürassier Regiment, the Leib Kürassier Regiment, Husar Regiments Nr. 8 and Nr. 11, Dragoner Regiment Nr. 8, Ulan Regiment Nr. 5, Kürassier Regiment Nr. 4, and Jäger zu Pferde Nr. 2 and Nr. 6. Other Garde regiments were also part of the Division, including 3. Garde Artillerie Regiment, Garde Maschinengewehr Abteilung, units of Minenwerfer troops, Kampfwagon Abteilung Körting, I. Marine-Brigade von Roden, and many others. They fought in the Berlin streetfighting December 1918, January 1919, and March 1919, moving on to Munich in April-May 1919. The saluting officer on horseback is in all probability Generalleutnant von Hoffmann. The marching officer on the left carries a P. 08 and an Imperial pilot's badge can be seen just above the holster. Three of the other four men wear their war-time Husar Attilas with the collar insignia of the GKSD and both M 1916 and M 1918 helmets.

OPPOSITE

Top: The band of the Garde Kavallerie Schützen Division marches ahead of a group of mounted troops. The band was formed from all units of the Division as can be seen from the mix of Infantry M 15 Bluse and Husar Attilas. Berlin, 1919.

Bottom: An ex-cavalry officer and his mount now in the Freiwilligen-Verband of the 31. Infanterie Division as indicated by his collar insignia. His visor cap in the high cavalry style, shows both cocards present and the Totenkopf cap badge of the Eisernen Division, and is constructed of two colors. The shoulder straps are light in shade and possibly woven silver thread of the wartime officer's type. Tall riding boots, spurs, and a dress bayonet, or fighting knife, round out his uniform for the day.

Left: An officer on the left and an enlisted man of the Freiwillige-Landeschützenkorps. Commanded by Generalmajor von Roeder and comprising of units of the Landeschützen Brigade and a flying Jagdstaffel. Photo is dated Oberhausen, 23.3.1919.

OPPOSITE

Top: Inspection of this photo with magnification shows two varieties of the oak-leaf collar insignia worn by these troops, but the Oberschlesisches Landjägerkorps is a positive identification choice. The man kneeling in the center has the Freikorps Oberland, or von Maden, Edelweiss pinned to his Mütze and most wear a yellow and white armband. The uniforms are M 1915 Bluse with one exception. The man second from the left is wearing a cotton stable jacket. He is equipped with leather puttees and laced boots rather than leg-wraps worn by most of the others. All weapons are Gew. 98s and the man standing fourth from the left has a long tobacco pipe over his shoulder.

Below: The Totenkopf flag of Freikorps Brüssow, or Freiwilligen-Corps Brüssow, as stated on the flag. The unit was formed by Leutnant Brüssow in Berlin, January 1919, and consisted of 1200 officers and men from infantry, machine gun and Minenwerfer specialties. Those looking on this snowy day include some Naval personnel, a drummer holding his drum with musician Swalbenesters on his shoulders, a boy with a sailor's cap, and various interested civilians.

Freiwilligen-Corps Brüssow
Für deutsches Land!

A postcard entitled on the reverse "Offizielle Reichswehr-Postkarte," "Reichswehr Verlag 1. Serie Nr. 6." Others from this series by the artist Sailer have not been encountered. The Freikorps Epp sleeve badge is clearly seen in this illustration.

OPPOSITE
When the Communist threat was at its highest, many Citizen's Defense Units, or "Einwohnerwehr", were formed and this recruiting poster was asking for volunteers and listed thirty-four registration locations in the Munich area.

Wehrmanns-
Ausweis
zugleich
Verpflichtungs-
und
Waffenschein

für den
Wehrmann
Karl Dürr

As with all military and para-military, the Einwohnerwehr members had an Ausweis, or identify card. This picture is the cover of the card issued to Wehrmann Dürr of Karbach in Bavaria.

Einwohner-
Wehr
Meldestellen:

Above: As uniforms were not issued to Einwohnerwehr members, those serving in the Munich area wore blue and white armbands with an identifying badge sewn to it. This style armband has been noted as being worn by uniformed troops without the badge.

Below: The inside of the Ausweis issued to Dürr in 1920 with his vital statistics, including his man number "43", as well as the number of his issued Gew. 98, 3825.

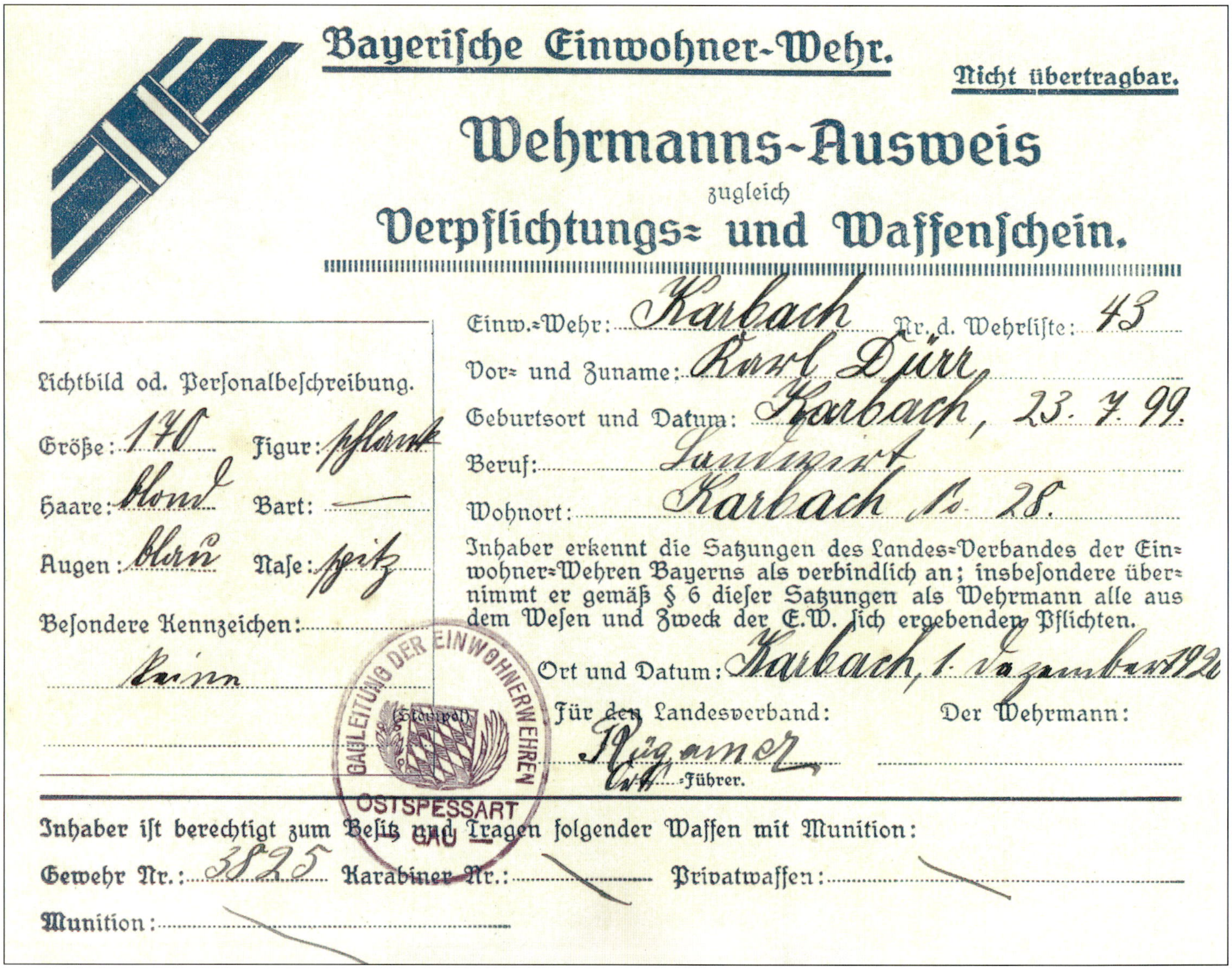

Bayerische Einwohner-Wehr.

Nicht übertragbar.

Wehrmanns-Ausweis
zugleich
Verpflichtungs- und Waffenschein.

Lichtbild od. Personalbeschreibung.

Größe: 170 Figur: schlank

Haare: blond Bart: —

Augen: blau Nase: spitz

Besondere Kennzeichen: keine

Einw.-Wehr: Karbach Nr. d. Wehrliste: 43

Vor- und Zuname: Karl Dürr

Geburtsort und Datum: Karbach, 23. 7. 99.

Beruf: Landwirt

Wohnort: Karbach No. 28.

Inhaber erkennt die Satzungen des Landes-Verbandes der Einwohner-Wehren Bayerns als verbindlich an; insbesondere übernimmt er gemäß § 6 dieser Satzungen als Wehrmann alle aus dem Wesen und Zweck der E.W. sich ergebenden Pflichten.

Ort und Datum: Karbach, 1. Dezember 1920

Für den Landesverband: Der Wehrmann:

Rügamer
Ort-Führer.

GAULEITUNG DER EINWOHNERWEHREN OSTSPESSART GAU

Inhaber ist berechtigt zum Besitz und Tragen folgender Waffen mit Munition:

Gewehr Nr.: 3825 Karabiner Nr.: — Privatwaffen: —

Munition: —

Instructions were given to members through this handbook which included rights and duties, as well as instructions for weapons storage.

Above: This photo gives a good illustration of the appearance of Einwohnerwehr troops. All but two are devoid of military garb, being dressed in their normal civilian clothes with only armbands to indicate their civil defense function. Those with visible weapons are armed with Gew. 98s. This group is from Einwohnerwehr, Duisberg.

Right: Oberst Reinhardt formed this unit in March 1919 and raised his forces in part through this simple recruiting ad which was placed in newspapers and weekly magazines. Their fighting was done in Berlin.

Below: Close-up of the first type Ehrhardt Brigade sleeve badge. The second type replaced the word "Wilhelmshaven" with "Ehrhardt" at the same location. The stitching to the black background can be plainly seen.

Above: A photo of an actual badge from Freikorps Bamberg which was formed in Bamberg 21, April 1919. A small unit consisting of five officers and 60 NCOs and men, they fought in Bamberg and Donauwörth, joining with Reichswehr Infanterie Regiment 46.

Below: Minenwerfer-detachments Heuschkel used another variation of the Totenkopf on their arm badge. The unit was formed by Hauptmann Heuschkel and consisted of eight light Minenwerfer sections, four heavy machine gun sections, one light machine gun section, and a flamethrower unit, totaling fourteen officers and 183 NCOs and men. Their fighting was contained to Berlin.

Above: A photo of members of Freiwilligenkorps Thüringen. The unit was formed in Thüringen where it fought in January 1919 in the city of Essen. Their strength was twenty officers and 500 NCOs and men. They disbanded 30 April 1920 and were absorbed by Reichswehr Infanterie Regiment 21 of Reichswehr Brigade 11. There is no reference to any insignia for this unit and none appear in this grainy picture. Uniform variations abound, with an M 1910 tunic, M 1915 Bluse, a partial uniform with a Naval cap and civilian clothes are all seen here. Small automatic pistols and Gew. 98s make up the visible armament.

OPPOSITE
A Freikorps veteran, H. Winkelmann, is photographed while serving with 2. Maschinengewehr Kompanie Reichswehr Regiment 16. The regimental number "16" is chainstitched on his shoulder straps, piped in white. On his left pocket is the Schlesischer Adler awarded for service in the Schlesien campaign. His uniform is the Reichswehr "Regulation", an improvement on the M 1915 Bluse with the addition of the two exposed breast pockets. His visor cap has the pinched look popular with long service soldiers.

Left: An NCO of the Landesschützen-korps. There is conflicting information on his insignia, one reference attributing the Mützen Abzeichen to the Freiwilligen-detachment Probstmayr; the other, which may be more reliable, states that the wreath and oak leaves are the insignia of the Landesschützen-korps. The commander of the Landesschützen-korps was Generalmajor von Roeder and the unit fought in the Berlin rioting and then three brigades fought in Bremen in 1919. Our subject wears an M 1915 Bluse, a waist belt with the Imperial crown supporting an M 98/05 bayonet with a Troddel identifying his company number. The reverse of the photo is dated Königsberg, 22.10.19.

OPPOSITE

Top: Four members of Freikorps Niedershein pose for an action shot. One man loads his Gew. 98, one man aims to fire, but the MG 08/15 crew won't be of much help without ammunition. The M 1915 Bluse on the left-hand man seems lighter in color than the other three. The contrast can be seen between the Bluse and the trousers on the man on the right. Freikorps Niedershein was commanded by Major von Schaumann and fought in the Ruhr campaign.

Bottom: Government troops (Regierungstruppen) fire a 10.5 Leichte Feldhaubitze on Alexanderplatz. Berlin, 8 March 1919. This artillery crew is not wearing steel helmets, which is unusual in the fighting at this time. Light field guns like this came in 7.7 cm, 10 cm and 10.5 cm with varying length barrels. Some designated "Kanone" and some "Haubitze" (cannons and howitzers), they were used sparingly during the fighting against the Communists.

ROSSBERG
erren u Damen
FRISEUR

OPPOSITE
Top: A Feldkanone C 96 m/A, 7.7 cm and its limber in the hands of government troops combating the Communists in Berlin, January 1919. The man on the right wears a long wide wool scarf over his Bluse; others wear the oak-leaf collar insignia of a Freikorps unit. In the background demonstrators carrying placards give the feeling of unrest of the times.

Bottom: A Feldkanone with painted wartime camouflage rolls through the streets of Munich, Spring 1919. The crew in steel helmets rides on the provided crew seats of the cannon, while others sit on the limber ahead of it. The boy on the right waves a small flag of blue and white Bavarian colors, while the rest of the crowd looks on happily.

Above: Württembergische Freikorps on the road to Munich, April 1919. Swastikas appear on their helmets and vehicles, before their adoption by the Ehrhardt Brigade. "Stuttgardt" and "Tubingen" are lettered on the side of the Krupp-Daimler Kw. 19, 100 horsepower heavy goods wagon towing the Feldkanone. One of the Kraftfahrer drivers leans on the spares box wearing an M 1912 Lederlitewka and a steel helmet over some other type of cap. The other troops in the rear are all wearing M 1915 Bluse with one exception who is also wearing the leather driver's coat with twelve front buttons.

Above: A photo postcard of a heavy Minenwerfer unit in action during the March Communist rioting in Berlin. The wicker ammunition transporting baskets gives one the sense of the size of each round. A round, ready to load, stands between the two men in the center. The crew, Government troops, not Freikorps, is armed with both a Gew. 98 and a Kar. 98a.

OPPOSITE

Top: A double-decked bus loaded with civilians passes a Government artillery unit supported by infantry during the fighting in Berlin, March 1919. The gun crew seems totally uninterested in the large crowd, held back by the infantry troops, gathered by the street corner.

Bottom: Government troops with a Russian Gardord-Patilow armored car on the Alexanderplatz. Berlin, March 1919. One of the many variations of the Totenkopf used by the Kokampf can be seen on the side and rear of the vehicle.

eisen
30
Fabrik

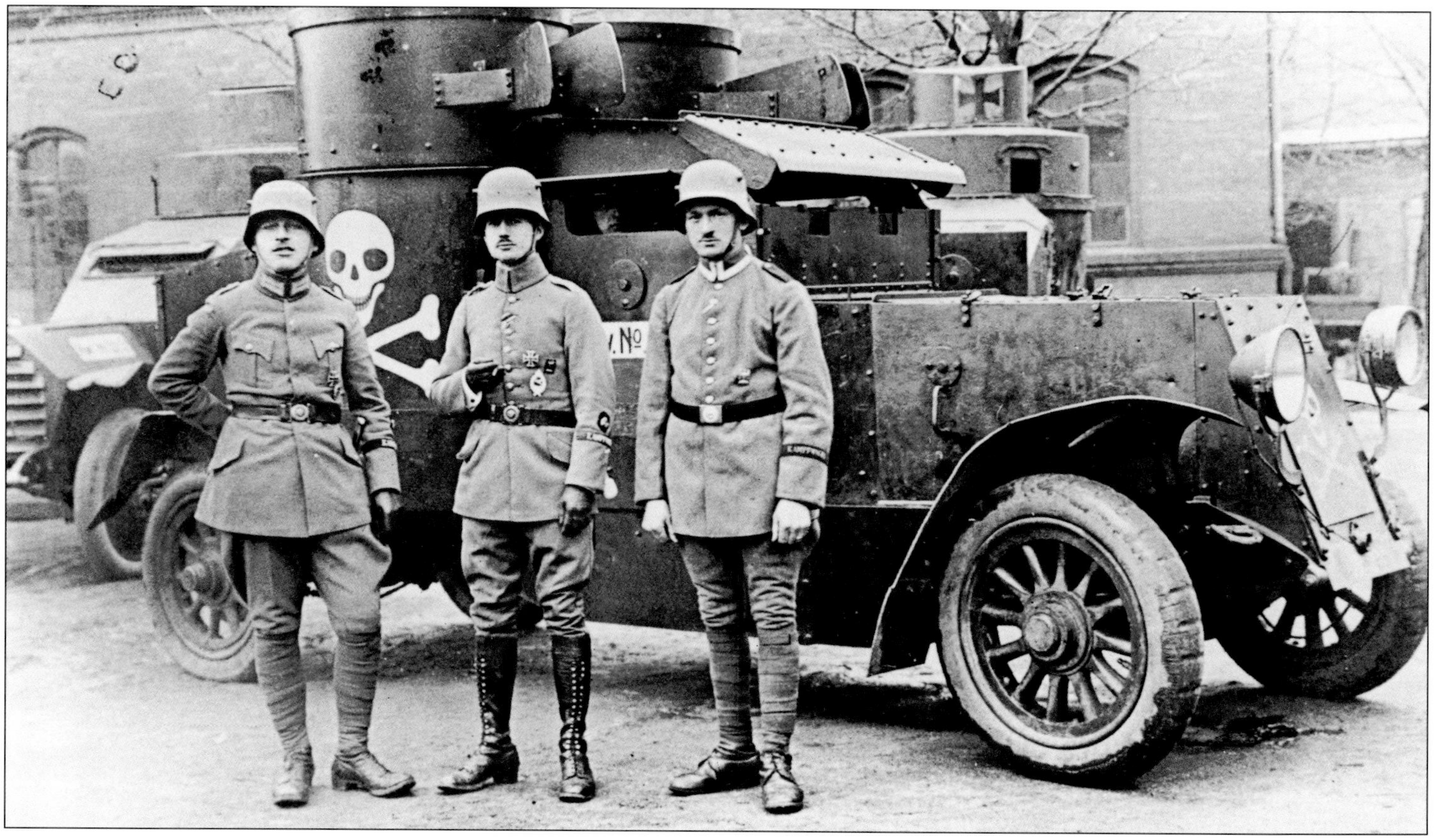

Above: Two Kokampf officers and an NCO stand by a Russian Austin of the unit at their Kaserne, Berlin-Lankwitz. The officer on the left wears a fly-fronted four pocket tunic with Litzen on the collar, an officer's leather Feldbinde, or belt, and on his tunic the Iron Cross First Class, a Wound Badge, the Kokampf's cuff title "Kampfwagen" and a small Totenkopf in a black circle above it. The officer in the center has the same decorations and in addition, the Flieger-Erinnerungs-Abzeichen, or flyer's commemorative badge, the Schlesischer Adler for Schlesischen service, and the ribbon for the Iron Cross Second Class. The NCO has rank tressing on the collar of his M 1915 Friedens-style uniform with plain cuffs and cuff title, but without the Totenkopf above it. He, too, has been awarded the Schlesischer Adler and another award shown by the indistinct ribbon in his buttonhole. Behind them is a captured Russian Austin armored car, and in the background, far left, a Lancia MK 2. This and the following picture are included by courtesy of the *Tank Museum Collection – Bovington, England.*

OPPOSITE

Top: The same Russian Austin, Kw. No. 024, which appeared in the previous photo, taken after a light snow at Berlin-Lankwitz. The vehicle was part of the Freiwilligen Kampfwagen-Abteilungen equipment. Close inspection shows the individual teeth that were painted on the Totenkopf insignia.

Bottom: An interesting group shot of Kokampf personnel, two in Naval uniform, as well as personnel from the Freikorps Eisernen Division as indicated by the Totenkopf insignia on their caps. The Naval Lieutenant zur See wears an assortment of Freikorps and Imperial badges. On his lapels are the oak-leaves of a Freikorps unit. Below his Imperial Iron Cross First Class is the Imperial pilot's badge for a Marine-Flugzuegführer. Between it and the Iron Cross is an unidentifiable circular badge. The white-capped Naval enlisted man has a more unusual combination. On the left sleeve of his jacket appear the following: the crossed anchors usually found in the insignia of one of the See-Bataillonen, or the Marine-Infanterie, units; below that the Totenkopf and cuff title "Kampfwagen" of the Kokampf. On his lapels, over his rank insignia, are the Freikorps oak-leaves. The white summer cover for his Naval Mütze is also unusual. Behind the Naval officer is a man wearing the motor driver's leather hat and Litewka with the Kraftfahrer insignia of the outline of an automobile on the wool collars. Between the two Naval personnel is an officer with Freikorps oak-leaves on his collar and high-polished boots. The armored car on the left of the photo is a Lancia with the Totenkopf insignia in black on a white circle and the vehicle is camouflaged. The car on the right is one of the Russian Fiat's. *Courtesy of the Tank Museum Collection – Bovington, England.*

Kw.Nº.024

Above: "Lotta" is a Russian modified Packard armored car of the Kokampf photographed in Munich during the fighting May 1-2, 1919. Heavy chicken wire covers the openings in the turrets to prevent damage by hand grenades. A white and blue Bavarian identifying pennant is flown from the left front of the car which was operating with no fenders.

OPPOSITE
Top: Another Lancia of the Kokampf with a different Totenkopf insignia on the turret. This shows the skull in profile with a scroll beneath. If there was a legend in the scroll, it cannot be seen in this photo. In the background is the 1917 Austin seen in previous illustrations. *Courtesy of the Tank Museum Collection – Bovington, England.*

Bottom: One of the Kokampf's captured British MK IVs in action during the Berlin fighting in January 1919. Behind it is a Wix Flammenwerfer crew and two men with an MG 08/15 on their shoulders. These are Government troops, not Freikorps. A civilian in dark clothes, a wide-brimmed dark hat, and a cane, wears a white armband with two unidentifiable marks on it, while he crosses the street in a casual manner.

Above: March 1919 Berlin streetfighting involves two Government "Strassenpanzerwagen" armored cars. The car on the left in an Austin and the one on the right is a Typ Fiat 1. version. Both are of Russian origin, captured during the fighting on the Eastern front.

OPPOSITE
Top: The Berlin streetfighting involved most of the armored vehicles owned by the Government's Kokampf group. This photo, which has seen greater exposure in print than many, shows the use of the visorless helmets and that of a captured Russian Packard armored car which is directly behind the helmeted troops. Two of these cars had been taken from the Russians and both were in Berlin in March 1919.

Bottom: Government troops huddle behind one of Kokampf's two British MK IV tanks during the Berlin streetfighting. This was, in all probability, Kokampf Nr. 153 named "Hanni". Nr. 154 was named "Fee".

M.K.III

Strassenkämpfe in Berlin
Kampfszene an der Volksbühne.

The same MK IV tank again with Government troops with the visorless helmets, M 1915 Mantels, and four-pocket Bluse.

Another view of the same tank from the other side, and in the background, an Ehrhardt E-V/4 is moving into position.

This photo shows a Daimler-Kraftwagengeschütz during the Munich streetfighting in early May 1919. The grille of the vehicle bears the Totenkopf of the Kokampf and carries eight heavily armed men. Not only is the vehicle armed with a cannon, but also an MG 08/15 facing to the front. The same or identical vehicle appears earlier in this volume.

Above: The presence of flower sprays held or worn by many in this picture would indicate a departure. Ritter von Epp, Commander of Freikorps von Epp, stands in the front row slightly to the left of center wearing a light-colored Mantel with flowers tucked in front. Many of the men are wearing the mid-1919 style of corded epaulets above the cuff rank insignia. The von Epp sleeve insignia is visible on the left arm of many, collar Litzen show on most of the uniforms, caps vary from bearing the Imperial-style double cocards to the Reichswehr-style with the lower cocard surrounded by a wreath. The man standing next to the train, holding a bouquet out towards von Epp, and the man next to him both wear the all new cloth Feldmütze bearing two cocards which came into being 5 May 1919. In September the regulation was changed and henceforth only one cocard was allowed. Freikorps von Epp fought in Munich and became part of Bavarian Reichswehr Brigade 21 and then continued to fight in the Ruhr. This departure, therefore, was taking place in Munich with the troops heading for the Ruhr.

OPPOSITE

Top: Another view of the well-photographed Daimler which at this point had an MG 08 added to the rear for additional firepower. The picture was taken at the Sendlinzer Tor, Munich, on 2 May 1919. Two of the men standing on the ground are wearing white and blue armbands in Bavarian colors indicating they are part of the Government forces.

Bottom: Berlin, 15 January 1919 Tank Abteiling Körting was operating this Whippet Tank, Kokampf "53", during the Berlin uprising. The tank had been captured and brought to the city for study in 1918. Seen here it was still in Berlin by the Eden Hotel which was the Headquarters for the Garde Kavallerie Schützen Division. It was experiencing engine trouble as indicated by the open motor access panels and was never seen again in action.

53

Above: This NCO acting as a chauffeur in Munich, May 1919, had served with General Lettow-Vorbeck in Ostafrika and was serving him again as a member of Freikorps Lettow-Vorbeck. The leather Mütze and the goggles are part of the equipment of a Kraftfahrer, or motor driver, but the uniform is a straight M 1915 Bluse with Garde Kavallerie Schützen Division insignia on the collars.

OPPOSITE

Top: A group of Freikorps von Epp from the Haushofer Zug, pose in the Kapuzinerstrasse. Munich, May 1919.

Bottom: Members of Freikorps von Epp pose before a foliated background. In this photograph we see an interesting variation of headgear: Dienstmütze visor caps with and without chinstraps worn without stiffeners, the new M 1919 Feldmütze with a cloth-covered visor and the old Imperial-style Kratzchen without a visor. All bear the double cocard which dates the picture earlier than late 1919. The von Epp sleeve badges with the lion's head are much in evidence, as are the M 1915 Bluse worn by all but the man on the far left who still favors the M 1910 style.

Above: A heavily armed detachment of Communist "Spartakus" revolutionaries pass through the Brandenburg Gate into Berlin, December/January 1918-19. Seen among the mixed weapons are an M 96 Mauser pistol, an artillery Model P. 08 pistol, an MG 08/15 light machine gun and Gew. 98s. The number of those in civilian clothes seems to be about equal to those dressed in uniform, or parts thereof. This view was printed as a photo postcard.

OPPOSITE

Top: The period caption on the reverse of this photocard reads "Republican Soldiers the end of November 1918". Most of the men are wearing M 1915 Bluse, many with NCO rank on the points of their collars and some still wear shoulder straps of a dark material. All the caps have only one cocard with the exception of the man kneeling in the center. This is an early photo of troops loyal to the Government.

Bottom: A goods wagon rolls through the Brandenburg Gate carrying a load of Communist Revolutionary troops. On the roof of the cab an MG 08/15 is in firing position, those in the back are armed with Gew. 98s, Kar. 98a's, and one man smoking a cigarette carries a Russian Nagant. The specific date is unknown, but it is late December/early January 1919.

Zimmer 44
genannt
Revolutionszimmer

Above: Berlin streetfighting. Loyal Government troops and armed civilians at the Brandenburg Gate. A post card, 1919.

OPPOSITE
Top: Loyal Government troops in position to repel the revolutionaries from atop the Brandenburg Gate. In the background one armed civilian joins the uniformed forces.

Bottom: This photograph, often seen cropped, has been on many occasions misidentified as being Spartakus, or Revolutionary, troops on the top of the Brandenburg Gate. These are loyal Government troops in a rather dramatic defensive position. Posed or not, it is a great picture.

Left: The original of this photo portrait is printed on a linen paper which makes it somewhat difficult to reproduce clearly. Of particular interest is the manner in which the Freikorps oak-leaf collar insignia are worn over strips of applied Iron Cross Second Class ribbon on the M 1915 Bluse. Close inspection reveals a small reserve cross on the lower State cocard on the visor cap, as well as the presence of shoulder straps on the Bluse.

OPPOSITE
Top: A group photo of members of the Bavarian Einwohrnerwehr. There are a variety of the blue and white armbands being worn which indicates there were several Einwohrnerwehr units involved. They are all armed with Gew. 98s which is the model of rifle usually encountered with these units.

Bottom: Streetfighting in Berlin, January 1919, with a 10.5 cm FK field gun and limber of the Government's loyal troops set for action. A Berlin policeman wearing a Tshako can be seen on the left and more "Regierungstruppen" are in the background.

Above: A captured Russian Austin armored car of the Panz. Kraft. Wag. Abt. is seen on the streets of Berlin/Neukölln in January 1919. The bemedaled officer-commander stands by the door wearing a fleece-lined coat, turtleneck sweater, and a neck award which looks like the Pour le Mèrite. On his officer's Feldbinde is a holstered eight-inch barreled P. 08 artillery model Luger and on his right arm, a white armband designating him as a member of the loyal Government troops. The turrets of the armored car hold MG 08/15s and its radiator armor plate bears the Totenkopf of the Kokampf, a Roman numeral "II" for Kampfwagenzug 2 is above it.

Above: A 7.7 cm FK 96 n/A manned by loyal Government troops blocks a street against revolutionaries in Berlin, January 1919. The warning sign tells the Spartakus to "stop and those who try to move on will be shot."

OPPOSITE

Top: General Hoffmann, Commander of the Garde Kavallerie Schützen Division, standing in a visor cap in front of the warning sign, stands with men of his division on the streets of Berlin, on 14 January 1919. A dachshund with a studded collar on a leash, is held by the P. 08 armed officer to the left of General Hoffmann. The men are wearing M 1915 Bluse and cavalry issue open-buckled belts. The warning sign makes the same statement as the one seen in the previous picture.

Bottom: Government troops man an MG 08 which is loaded and ready to fire on a wide street in Berlin on 11 January 1919. This photo and others taken about the same time show a wet and chilled Berlin during this stage of the fighting.

Halt !

Sachs, Obert & Co.
2. Hof. Aufgang 4
Eckart Verlags

Above: A "Wachmannschaft" of the Garde Kavallerie Schützen Division eating and enjoying refreshments at the Eden Hotel, the Division's headquarters during the rioting. Seated in the center of the table is the not too bright and notorious Soldat Otto Runge who on 15 January, acting on orders, clubbed both captured Spartakus leaders, Karl Liebknecht and Rosa Luxemburg, close to death as they were being transported from the Eden Hotel. He later confessed "...I had strict orders from the officers to club those scoundrels to death with my rifle butt as soon as they came out". For this he got two years in prison after being dismissed from the Freikorps for "leaving his post without being properly relieved" and "improper use of his weapons". This photograph bears the date of 15 January 1919 and it is unknown if it was before or after the murders.

OPPOSITE

Top: A captured Russian Austin armored car named "Leni" at the Berlin-Lankwitz Kokampf "car wash." The Totenkopf insignia is on the front, as is a single headlight in the center of the hood, rather than a pair down by the fenders. The man with the hose is dressed in a Kraftfahrer leather coat and visor cap.

Bottom: Members of Kampfwagenzug III of the Kokampf pose by two Daimler DZVRs and an unidentified touring car. The firepower potential of these DZVRs can be seen by the number of MG 08/15s bristling from all sides. "Gerda," the car in the background, has a small forward hatch open revealing a single spotlight on the turret. The crews in the foreground wear a wonderful assortment of dress for armored car service. There are four examples of the M 1912 Lederlitewka, M 1915 Bluse, and the Panzerkraftwagen one-piece coveralls. Headgear varies from visor caps to leather crash helmets, to an aviator-style helmet to leather skullcaps worn by the two men, far right.

Leni

Gerda

Above: Members of Kokampf's heavy detachment, May 1919, surround one of their captured British MK IVs and "Hedi," the modified A7V. "Hedi's" awesome firepower can be seen by four visible MG 08/15s with others not in view. Standing on the forward part of "Hedi" is a man displaying the Kokampf flag of a black Totenkopf on a black, white, and red national flag. Seen on many of the uniforms are Freikorps collar insignia, "Kampfwagen" cuff titles, Kokampf white Totenkopf insignia on black circles, machine gunner's badges, Iron Cross First Class and Imperial pilot's badges. The officer marked with the double "X" is Ltn. Theunissen and with the triple "X", Ltn. Bornschien. Rail Station, Leipzig, May 1919.

Left: Unfortunately this poor quality photo is the only known close-up of Oberleutnant zur See Gotthard Sachsenberg, Pour le Mèrite Naval flyer with thirty-one victories. He commanded Marine Jagdgruppe I in Flanders from September 1918 to the end of hostilities, and subsequently formed Freikorps Kampfgeschwader Sachsenberg which flew in the Baltic operations.

OPPOSITE
Top: Kampfgeschwader Sachsenberg officers pose for a photograph before departure for the Baltic. Oberltn. Zur See Sachsenberg wearing his Pour le Mèrite links arms with Oberleutnant zur See Eberhard Cranz who wears a blue Naval officer's reefer jacket and a bow-tie. Another Naval officer identified by his cap is behind Sachsenberg and another is on the far right. The unit consisted of fifty officers, 650 men and seventy aircraft, making it a sizable unit.

Bottom: Leutnant Carl August von Schoenebeck on the left, stands with an unidentified friend in front of a Siemans Schuckert D. III A in this postwar photo. The Siemans Schuckert, although not built in quantity, was considered by many the best fighter aircraft in World War I. For some reason, Schoenebeck is not wearing his pilot's badge on his uniform. He went on to fly with the Freikorps in the Baltic and was wounded at Mitau in 1919. Previous to Freikorps service, he ended World War I as C.O. of Jasta 33, with a total of eight victories.

Above: For a period before the signing of the Treaty of Versailles, the provisional Reichswehr maintained parts of the Imperial cavalry forces which were, after the Treaty absorbed into the Reichsheer, forming 18 cavalry units, Reiter-Regiments 1-18, which did not continue the use of the specialized uniforms as worn by Husaren, Ulanen, and Kürassieren regiments in the Alte Armee. These Ulan musicians are not wearing the traditional Schwalbenester shoulder devices worn by all musicians on foot or on horseback, but do retain their cloth-covered Tschapkas and Ulanka tunics. A hornist in the second rank wears the three-bar rank insignia of a Vicefeldwebel as dictated by the regulations of early 1919. This picture is taken in Munich after the May Revolutionary riots and these troops are from Bavarian Ulanen-Regiment 1 or 2.

OPPOSITE

Top: Light Minenwerfers normally were served by crews of four, but it would appear that Freikorps and Government troops during the revolution cut it down to three. Here are two crews of three with officers and infantry in support. Munich, May 1919.

Bottom: An "MG-Gruppe" of the 1. Bataillon Freikorps Oberland. This "Gruppe" wears a great collection of Freikorps badges. Their caps show not only the Totenkopf of the Eisernen Division, but also the Edelweiss of the Freikorps Oberland and both single and double cocards. Collar badges worn in this photo are the Edelweiss of Freikorps Oberland and most wear the machine gunners badge. The kneeling man wearing a shirt and tie under his M 1915 Bluse has an armelabzeichen in the shape of a shield, designating 1st or 2nd company of the first Bataillon, plus a Totenkopf below it. The man second from the right wears a forester's style soft felt hat and an M 1915 Bluse. Also of interest is the seated youngster in front of the MG 08 with its cleaning rod across his knees.

München, Mai 1919

OPPOSITE
Top: A Fahnengruppe of the Freikorps Oberland. The Edelweiss collar insignia can be seen on their uniforms which are all M 1915 Bluse with the exception of the man on the left who is in an M 1910 tunic and is armed with a Kar. 98a. A P. 08 in its holster is on the belt of the man with the flag and an unidentified pistol is carried by the man next to him. The badge on the sleeve of the man on the right is too indistinct to identify positively.

Bottom: These two bicycle dispatch riders belong to the Kompanie Oestreicher created and commanded by Ludwig Oestreicher on 21 May 1919 and part of Freikorps Oberland. Their unique arm insignia was an oval wreath of leaves topped by a crown and within the wreath, a rampant lion and the shield of Bavaria.

Right: Otto Dietrich was a member of the 4th Company of the 1st Batallion of the Freikorps Oberland. Dietrich was a man who enjoyed wearing insignia as indicated by this photograph. There are eight separate insignia shown: an Edelweiss on his cap for the Freikorps Oberland, another on his left collar for the same purpose, a Totenkopf on his right collar, also worn by other members of Oberland, a sash with some legend embroidered on it, the Schlesische Adler 2nd Class, under which is the badge for the Selbstschützes Schlesian. On his arm is the badge for 4. KP./1. Btl. Freikorps Oberland, (each company 1-4 had its own badge) and lastly, the machine gunners badge. There are Bavarian buttons on his tunic and he wears a cavalry-style roller buckle belt.

Right: Berlin photographer Willy Römer took this photograph of Gustav Noske in 1919. Noske was one of the most important figures during the uprisings from December 1918 through 1919. He was an ex-serving soldier attaining NCO status under von Moltke before entering politics and Government. At the time of the Kiel Naval mutinies, he was Governor of Kiel and helped control that situation after which President Ebert and General Groener decided he was the man for the Supreme Commander of the armed forces and Defense Minister. He was a great friend and promoter of the Freikorps movement and did all within his power to make the movement a much needed success. He aided in the removal of the Spartakus plague in Berlin and Munich and by holding back all loyal troops from firing on those involved in the Kapp-Lüttwitz Putsch, did much to further good for the overall Government cause. Noske later resigned as Minister of Defense and became the Minister-President of the province of Hanover, a position he held until forced to resign by the Nazi's. During World War II he was arrested twice by the Nazi's, the second time for a perceived association in the General's revolt and plot to kill Hitler.

Below: After the heavy fighting and the defeat of the Revolutionary forces in Munich, fighting broke out in the Ruhr. This photograph shows members of Freikorps Werdenfels, still wearing armbands and civilian clothes in lieu of uniforms, on a train headed for the Ruhr.

On 12 March 1920, an unusual event took place in Berlin – an almost comic opera performance called the Kapp-Lüttwitz Putsch. Webster describes a "putsch" as "a secretly plotted and suddenly executed attempt to overthrow a government" which indeed the Kapp-Lüttwitz Putsch attempted to do. Several previous events pushed these two men and others to attempt to take down the Republican government and replace it with one more friendly to the military and actively defiant to the dictates of the Versailles Treaty. The Versailles Treaty which had been signed by the Ebert Government was considered "a stab in the back" to Germany and particularly to the Armed Forces. There was still a very strong Imperial feeling among many of the military leaders, as well as an extreme dislike for the Republican government and its political positions. For years Wolfgang Kapp had been an obscure civil service employee in charge of the East Prussian Land Office. In the spring of 1916 he gained national recognition by founding the Fatherland party. Kapp had no visions of becoming the "new strong man of Germany", as he felt the position should be held by an Imperial general. Kapp thought Field Marshal Ludendorff would be a good choice, but in reality Ludendorff was too unpopular with the people due to his ruthless dictatorial war-time excesses. Others were approached, and as his last choice, General Freiherr von Lüttwitz, was offered the job.

Another event which infuriated General Lüttwitz and the other plotters was the planned attempt by the government to disband and demobilize the "Volunteer Army" which was, in essence, the Freikorps. Those members of the "Nationale Vereinigung", the new name for Kapp's Fatherland party, were alarmed at this prospect as they thought the Freikorps' last chance to seize the government was going to be denied. The Ehrhardt Brigade, some 5,000 strong, was then stationed at the comfortable barracks at Döberitz when the word came that as a Freikorps unit they were about to be disbanded. Ehrhardt immediately went to General Lüttwitz in Berlin and told him that his men were angry and ready to attack the City. Lüttwitz assured him that disbandment was not to happen and Ehrhardt hurried back to Döberitz to calm the troops. On 1 March 1920, the Second Marine Brigade (Ehrhardt) celebrated its first birthday by holding a special parade reviewed by Lüttwitz himself.

Below: The Kapp Putsch. Troops from the Freikorps Ehrhardt Marine Division roll into Berlin with an Imperial Naval battle flag flying. The helmets worn by Ehrhardt's men had large white swastikas painted on the front in a straight up and down position. The truck, with its front wheels clad in steel bands with springs underneath, necessary due to the war-time rubber shortage, is well-armed with an MG 08/15 on top of the cab and five Bz. 15 grenades hanging next to the driver. Two of "Berlin's finest" in their dark blue police uniforms seem to be unconcerned about the events unfolding.

On March 12, the Putsch began and they marched on Berlin, passing through the Brandenburg Gate at 7:00 AM and by noon Berlin was theirs. Berlin was taken without firing a shot. The Freikorps had done their job. Now it was up to the politicians. The politicians lack of preparations as how to assume office, and what to do in office, plus the advent of the General Strike called by the Social Democrats as they fled Berlin, caused the failure of the Putsch. The strike was put into effect in a concerted way by the workers of Germany and spread widely throughout the country, but was most effective in totally shutting down Berlin. As a result there was no water, no lights, no gas, no municipal transportation, no garbage removal and a citizenry with mounting anger and hostility against the new government. On the afternoon of March 17, a sullen Ehrhardt Brigade, on orders, left the city, marching past a hostile crowd of citizens. In the crowd a small boy laughed seemingly at the troops, whereby two soldiers broke rank and beat the boy to death. No one made a move to stop the violence, but someone made a hissing sound at which point an officer gave a command which commenced a burst of machine gun fire into the civilians. The Brigade then quickly reformed ranks and silently, in impeccable order, marched out of Berlin through the Brandenburg Gate, back to Döberitz.

Above: The same truck with Ehrhardt Brigade troops has reached Potsdamer Platz. Faintly the words "Mar. Brig. Wagen" can be seen on the sides, as well as a backwards swastika on the radiator indicating this was not a vehicle quickly requisitioned just for the entry into Berlin.

OPPOSITE
Top: A smaller truck than the one previously seen bringing Ehrhardt Brigade troops into Berlin during the Kapp Putsch. This vehicle has a heavy MG 08 mounted on the cab and the men are wearing vertical swastikas on their helmets. An M 18 cutout helmet has been issued to the man sitting on the fender. The officer behind him can be identified as such by the officer-style shoulder boards on his overcoat. Citizens in the street seem to be warmly welcoming the anti-government soldiers.

Bottom: The same truck surrounded by civilians at another location in Berlin. The officer seen before now stands in the street in the light helmet with his back to the crowd. Behind the military truck is one of civilian nature providing transportation for the populace.

IA-1425

Damen Konfektion

Right: Propaganda is being passed out by the Ehrhardt troops wearing helmets bearing swastikas of varying designs. The man in the foreground is dressed in an M 1915 Bluse with Litzen on the collar and the Ehrhardt Brigade arm badge. He is armed with an M 96 Mauser automatic pistol in its wooden shoulder stock holster, the usual grenade, M 1909 Patronentasche and a bandolier with spare rifle ammunition. The dapper man in the fur-collared coat reads the propaganda with interest. Berlin, March 1920.

Below: A most unusual unit designation is clearly seen on the door of the touring car and faintly on the front of the helmets worn by the men in both vehicles. The design is without doubt a stylized question mark, but there is no known explanation as to its meaning. This truck looks a bit weary and well used, but certainly to be contended with as shown by the sled-mount MG 08 mounted on the cab.

Right. A Mercedes staff car bearing the Imperial eagle becomes a platform for an MG 08/15 during the Kapp Putsch, Berlin, March 1920. The location is the corner of the Unter den Linden. Note the wing collar and tie worn by the central figure which is of the style worn by Naval officers.

Below: Serving with the Marine-Brigade Ehrhardt were Freikorps units which had fought in the Baltic. They had suffered an ignoble defeat at the hands of the Government politicians who called an end to hostilities rather than victory by the Poles and Communists. Bitter, they eagerly joined in with the attempted Putsch to avenge themselves from the Government in power. This heavy machine gun section are wearing the sleeve chevron of the Lebtstschütz-Bataillon Marienburg, formed by Kapitänltn. Lensch. Practically every man carries an ammunition box, as well as many with dragging straps, Kar. 98a's, a P. 08, Bz. 15s, and full packs. One cutout style helmet is seen, but there seems to be no insignia painted on it, or any others in the photo.

Above: Opposing Generals meet during the Kapp Putsch. General von Lüttwitz an arrogant and not too brilliant commander on the left, commands the Marine Brigade and other putschist forces. General von Hülsen, commander of Freikorps von Hülsen, by direction of Noske takes a non-aggressive position and refuses to engage von Lüttwitz's revolutionaries, therefore making the attempt bloodless.

Above right: Ehrhardt's Marine Division raised the Imperial Naval battle flag over Berlin during the first day of the Kapp Putsch. The man on the right wears an M 1915 Bluse, Infanteriestiefel, or marching boots, and is armed with a Mauser M 96 pistol in its wooden shoulder stock/holster with the leather carrying rig. Under the M 96 there is what looks like a short Naval cutlass, and two grenades complete his armament. Under magnification a swastika is painted on the helmet of the man on the left, as well as Garde Kavallerie Schützen Division insignia on his collar.

Right: Hanging from the small carriage for the transportation of light MG 08/15s and covered by equipment, is an MP 18 with its snail drum magazine attached. Another of these scarce MP 18s shows in an earlier photo of Ritter von Epp inspecting his troops. Waffenfabrik Theodor Bergmann/Suhl was the manufacturer. After April 1918, the weapon was known officially as MP 18, I. The weapon used 9 mm ammunition, the same as the P. 08, fed to this gun by a Stangenmagazin of 20 rounds, or by a Trommelmagazine, called by some a snail drum magazine of 32 rounds. The carriages for the other MG 08/15s, plus stacked Gew. 98s, are in the background. The Imperial Naval battle flag flies proudly on the left.

Above: Berlin, March 1920. The Ehrhardt Naval Brigade controls the city and the troops are awaiting further orders. In this street scene, we see stacked Gew. 98s, M 1914 Ersatz Tornister packs, M 98/05 and Seitengewer 14 typ II bayonets, M 1915 Bluse with Ehrhardt Brigade sleeve badges, M 1918 helmets and the Imperial Naval battle flag prominently displayed.

Below: A picture of the same display of equipment taken from a different position. The personnel are different and some interesting equipment is worn by them. The second tall man from the right has two snail drum magazines tucked in his belt and the Ehrhardt badge on his sleeve, the Garde Kavallerie Schützen devices on his collars, a Wound Badge and a pair of small single leather pouches on his belt. Two men, further on the left, have similar pouches.

Above: Seated in a staff car is Capitain zur See Ehrhardt himself and some of his staff in conference with a well-dressed civilian. Ehrhardt has a set of heavy woven epaulets on his overcoat indicating his Naval rank. The officer on the right wears the collar badges of the Garde Kavallerie Schützen Division (G.K.S.D.).

OPPOSITE

Top: Ehrhardt's Naval Brigade lines up for review. All are wearing M 1915 Bluse with G. K.S.D. collar badges. All but the man on the left who is armed with a Kar. 98a, are armed with Gew. 98s. The man on the far left is dressed in Sturmtrupp garb, steel helmet, M 1915 Bluse, trousers reinforced with leather at the knees, goat hair high socks and laced boots. On his collars are the G.K.S.D. insignia coupled with, on his chest, the badge of the Deutsche Schützdivision, which was a sword through a round wreath.

Bottom: A different view of the same formation. Ehrhardt Marine Division sleeve badges are much in evidence, as are the collar badges for the Garde Kavallerie Schützen Division. Several of the men in the left-hand rank are equipped with small leather single ammunition pouches which have been noted in an earlier photograph. No reference has been found to identify these pouches.

II-Aa 7

Photothek Berlin

Truppen der Regierung Kapp a.d. Potsdamer-Platz.

Above: Another artillery piece of the well-equipped Kapp forces, a Krupp 10.5 cm LFH is still in wartime camouflage. Berlin, March 1920.

OPPOSITE

Top: Mixed uniforms were worn by the forces of General Lüttwitz while occupying Berlin, March 1920. The first man on the left wore the M 1915 Mantel, steel helmet, straight trousers with no leg-wraps, and was equipped with a Gew. 98 and an Ersatz bayonet. The next man, an NCO as indicated by his round collar disc and NCO tressing, wore an M 1910 tunic and a visor cap. Next to him are men in the M 1915 Bluse, one of whom has a swastika painted on his helmet. All were equipped with M 98/05 bayonets.

Bottom: The caption on this photo postcard identifies these to be troops of the "Kapp Government" in Potsdamer Platz. Interesting that they were called "troops of the Kapp government" which certainly was a short-lived one. The gun numbered "3" is a 7.7 cm FK 96 n/A and is identified as being from the Michaelis Btn. Wicker baskets protect the 7.7 cm rounds stacked by the trail of the gun.

Right: Kapitan Ehrhardt salutes his troops on their entry into Berlin, March 1920. To his right is one of his decorated officers wearing the Iron Cross First Class, silver Wound Badge, the square Verdienstabzeichen of Marine Brigade Ehrhardt, the Naval Brigade Badge on his upper sleeve and above the cuff, an unidentified insignia. To Ehrhardt's left is another officer with the Iron Cross and the Garde Kavallerie Schützen Division insignia on his collar.

Above: A machine gun detachment of the Marine Brigade wearing the Ehrhardt arm shields, move their carriage mounted light MG 08/15s through the Potsdamer Platz in Berlin. Some of these troops also wear the collar badges of the Garde Kavallerie Schützen Division.

OPPOSITE

Top: Even though this picture is of poor quality, the carriage mounting system for the MG 08/15 can be more clearly seen.

Bottom: The Marine Brigade takes a break near the Brandenburg Tor, Berlin, March 1920. Battle packs, stacked rifles, Naval battle flags, a drum, and many Ehrhardt Brigade badges can be seen in this photograph.

Above: Compare this with the preceding picture. The drum and packs are in the same position, but the stacked rifles have been removed. One wonders which photo was taken first.

OPPOSITE

Top: It is difficult to determine what was happening when this picture was taken. A musical band of troops in Tshakos rests in the center, troops "at ease" in the foreground, horse-drawn equipment on the right and masses of troops in the background all must have some purpose on this rainy day of revolution in Berlin.

Bottom left: Marine Brigade troops are led by an officer on a white horse wearing the Ehrhardt badge, the Garde Kavallerie Schützen collar insignia, and the Iron Cross First Class. The Naval flag carries streamers from its pole-top and the pole has a spiral decoration on its lower section. These troops look as fresh, smartly dressed, and well-disciplined as any regulars could be.

Bottom right: Kapitan Ehrhardt and staff waiting for a review of his troops. The officer on the right can be seen more clearly than in a preceding photograph. The Garde Kavallerie Schützen Division collar badges, Iron Cross First Class, Wound Badge, the square-shaped Ehrhardt Verdienst badge are all in sequence. The Viendienst badge was partially cutout in the center and backed by a black, white, and red ribbon. In the center was a round wreath above which, in a rectangle, were cast the words "Verdienst ums" and below "vaterland." The cutout shape of the Viking ship on the Ehrhardt badge can be seen on the center officer's sleeve.

Ehrhardt.

Kapp's forces had an armored train in their arsenal of weapons, shown here and in the following picture. It was equipped with guns in revolving turrets, 10.5 cm gun on an open flatbed and MG 08s firing from armor plated cars. The sides were painted with the shield of the Teutonic Knights Order widely used by the Eastern Frontier Defense Forces. The same officer in a light Mantel and a brocade officer's belt appears in both photos.

Above: The exact date of this photograph has not been determined, but 1920 is the year. General Lüttwitz is making awards to his troops at the military Camp Zossen, which was 35 miles from Berlin. This presentation was probably made at a time prior to the Kapp Putsch after which General Lüttwitz fled to Hungary having completely fallen from favor as a result of his involvement and bad handling of the affair. Cutout helmets are worn by two of the awardees who also wear the collar insignia of the Garde Kavallerie Schützen Division.

Below: The extent of the equipment of Ehrhardt's Naval Brigade is indicated by the unit's portable radio wireless communications section. The portable generator is on the left on the ground in front of its transport wagon, the transmitter/receiver on the right in its wheeled transport. The sleeve badge of the Ehrhardt Brigade, as well as the Garde Kavallerie Schützen Division collar insignia, is worn by the operator on the right. Is lunch being served by the men in the center of the photo? No cook fire is evident and the pot is hanging from a frame with only two legs supported by the man on the right.

HERTA
S.Kw.A.10

OPPOSITE
Top: The fighting broke out again in the Ruhr after the failure of the Kapp Putsch and units of the Kokampf responded. This unfortunately fuzzy photo is of a Daimler DZVR 1919 of Schwer Kampfwagen Abteilung 10 (S. Kw.A10) taken in Wessel. The great size of the armored car can be appreciated by its comparison to the crew. The name "Herta", a Totenkopf, an Imperial Maltese cross, plus the name of the unit are painted on the side of the car.

Bottom: Another view of a Daimler DZVR gives a complete side view, including the turret with its forward mounted armored searchlight. The scale and size in this picture cannot be appreciated without a man beside it for comparison.

Above: The last tank in Reichswehr service was a captured British MK IV operated by Kampfwagen-Zug 13 of the Kraftfahr-Abteilung of Reichswehr Brigade 13. After the Versailles Treaty forbade the use of tanks by the Reichswehr, this one was hidden from the Allied Commission inspection at a remote artillery Kaserne near Stuttgart. The men in this photo are in Reichswehr uniforms. The small corded shoulder straps, the oval unit sleeve identification, and upper pockets give evidence to the fact. Some are still showing the Imperial machine gunners badges and one man 6th from the left has very large light colored chevrons on both sleeves. This photograph is from the Hauptstaats-archiv, Stuttgart.

Above: Freiwilligen Abteilung Schlesien of the Selbstschutz Oberschlesien. The three men on the left are dressed in similar tan colored jackets with shirts and ties underneath. Four wear the yellow/gold and white armband of the Selbstschutz low on their sleeves, and all are showing the Selbstschutz Oberschlesien stamped metal shield above the armband. The shield design consisted of a vertical sword over coiled snakes. The two men on the right have the look of White Russians who have joined the German's against the Reds. Brown leather belts with cross-straps are worn by all, the second man from the right has the cross-strap over his left shoulder, the others the right. The center man was awarded the Prussian flyer's observer's badge, Wound Badge and another decoration (Iron Cross First Class?) partially hidden by his bayonet.

A close-up photo of the Selbstschutz armband.

Above: This picture and the next are from von Salomon's history of the Freikorps and included not because of their clarity, but because of the unusual subject matter. The MG 08 crew have their gun operating on a trench mount rather than the regular sled mount. The gun has a strange long barrel projecting from the water cooling jacket, a 250-round ammunition box rests in the foreground. The unit is identified as Kompanie-Schlageter from Sturmabteiling Heinz. The kneeling observer, binoculars in hand, can be well identified by his arm insignia on the top, the Armelazbzeichen of Selbstschutz-Sturm-Abteilung Heinz, next, the single bar Reichswehr rank stripe for a Gefreiter, and finally, a machine gunner's badge. This unit decorated their equipment and helmets with vertical left-facing swastikas.

Above: This poor quality photograph has been reproduced in several works on the Freikorps and originated from von Salomon's book. It depicts an officer and an enlisted man of Selbstschutz Sturmabteilung Heinz wearing the insignia of the unit. Officers wore the badge on the lower sleeve, while enlisted men wore theirs on the middle part of the sleeve. The officer is a Leutnant as indicated by the single Reichswehr rank stripe. Both show the units' stamped shield on their breasts.

Left: A close-up photo of the Heinz arm badge, with yellow, white, and yellow ribbon.

While the Freikorps were putting out Revolutionary fires, the Provisional Reichswehr was hard at work adopting the new role of a Versaille Treaty mandated 100,000 man army. Under the control of Generaloberst Hans von Seeckt, a strong and highly disciplined army was planned. Gen. von Seeckt had no patience with the free-booting style of much of the Freikorps and as a result wanted little of their help towards his aim. Called the "Sphinx with a Monocle", von Seeckt was not fond of the Republican Government, but he was willing to bide his time in an orderly un-revolutionary manner to build a carefully selected, rigidly disciplined and diligently trained army of 100,000 "officers". When the Treaty could be circumvented, the army then could grow quickly with a huge trained officer cadre ready and in position to lead.

Above: This 1920 photo entitled "The Harmless" from "Room 53" shows a group of above average appearing soldiers in a posed off-duty scene. The Reichswehr oval unit identification patch, or Armspiegel on both sleeves, is for the 41st Infantry Regiment. Veterans of the Freikorps have the von Epp insignia below the unit number. The wartime M 1915 Bluse with added collar Litzen and narrow shoulder straps, is worn by all but two, one of whom on the right is in an M 1910 tunic. The visor cap, or Dienstmütze, still has two cocards, the lower surrounded by the Reichswehr-style wreath. One man has an Edelweiss on his cap which is a carryover from his Freikorps Oberland service. A 98/05 bayonet protrudes from a half loaf of dark bread; the other half to the left has a sharper knife for cutting it.

OPPOSITE

Top: This photograph shows the variety of dress worn by the Reichswehr in its infancy. The Armspiegel oval is numbered "41" and the Freikorps von Epp arm badge is below it. Shoulder straps when worn are of the narrow corded style first used in the early 1919 regulations. The uniforms include an M 1910 tunic on the man seated lower left, M 1915 Bluse on three in the back row, M 1919 fly-fronted "Regulation" model tunic with two breast pockets on the man seated far right, and finally, the two center seated men wear a modification of the M 1915 Friedens tunic which has exposed buttons, no breast pockets, unadorned cuffs, collar Litzen and narrow corded shoulder straps called Schulterschnur. In one instance a single chevron rank insignia is seen between the Armspiegel and the von Epp badge indicating an unteroffizier.

Bottom: The entire caption on this photo cannot be seen, but it probably has to do with "a Sunday without work" or "duty." This group all wear the M 1915 Bluse with oval Armspiegel indicating the 41st Regiment. Some are veterans of Freikorps von Epp and wear that badge also. A variety of head gear include the new Feldmütze M 1919 with the cloth-covered visor, the M 1919 Dienstmütze with single cocard and Reichswehr wreath and eagle. The Feldmütze has a colored band, plus an unidentified badge on it. The standard regulation M 1919 Feldmütze did not carry a colored band. In the far back left an earlier style Kratzchen, or Mütze, is still in use. Leather reinforced trousers and goat hair socks give the man on the right a storm trooper look.

Ein Sontag ohne

Left: The new Reichswehr uniform as issued. The Feldrock has green cuffs, two breast pockets and Litzen on the collar. The Dienstmütze is a high-crowned Imperial style without a chinstrap decorated with a black, white and red cocard above a wreath surrounding a drooped winged black eagle in an oval. An open buckle belt Schlosschnelle, rather than the rectangular solid "Gott mit Uns" type Koppelschloss is worn in this walking-out uniform.

Right: An "Alte Freikorpskampfer" from Freikorps von Epp serving now with the Reichswehr. The specific unit cannot be identified from his Armspiegel. The first model Dienstmütze, M 1915 Bluse, leather reinforced trousers, Ledergamaschen puttees, and hobnailed laced boots make up the uniform of this soldier who from his dress would not necessarily be from the cavalry branch.

Below: Youthful Reichswehr recruits "ham it up" for a group photograph with two older men standing on the right. Interesting uniform and equipment items appear in the picture: white undershirts with a colored band around the neck, buttoned front work jacket of cotton (drillichanzug), Kratzchen with the Reichscocard without wreath on the band, Dienstmützen with only the Reichscocard within a wreath and no upper cocard. In the back row the four right hand men are wearing the 1919-20 Feldrock with light pebbled buttons and two breast pockets. This photo was taken after the 1921 uniform change which among other things brought back the numbered wider shoulder straps and did away with the numbered Armspiegel. Leather tall marching boots, Stiefel, were replacing woolen leg-wraps. M 98/05 bayonets and Spaten M 1898 in both the 1898 and the M 1916 Spatentasche, or carriers, are all in evidence.

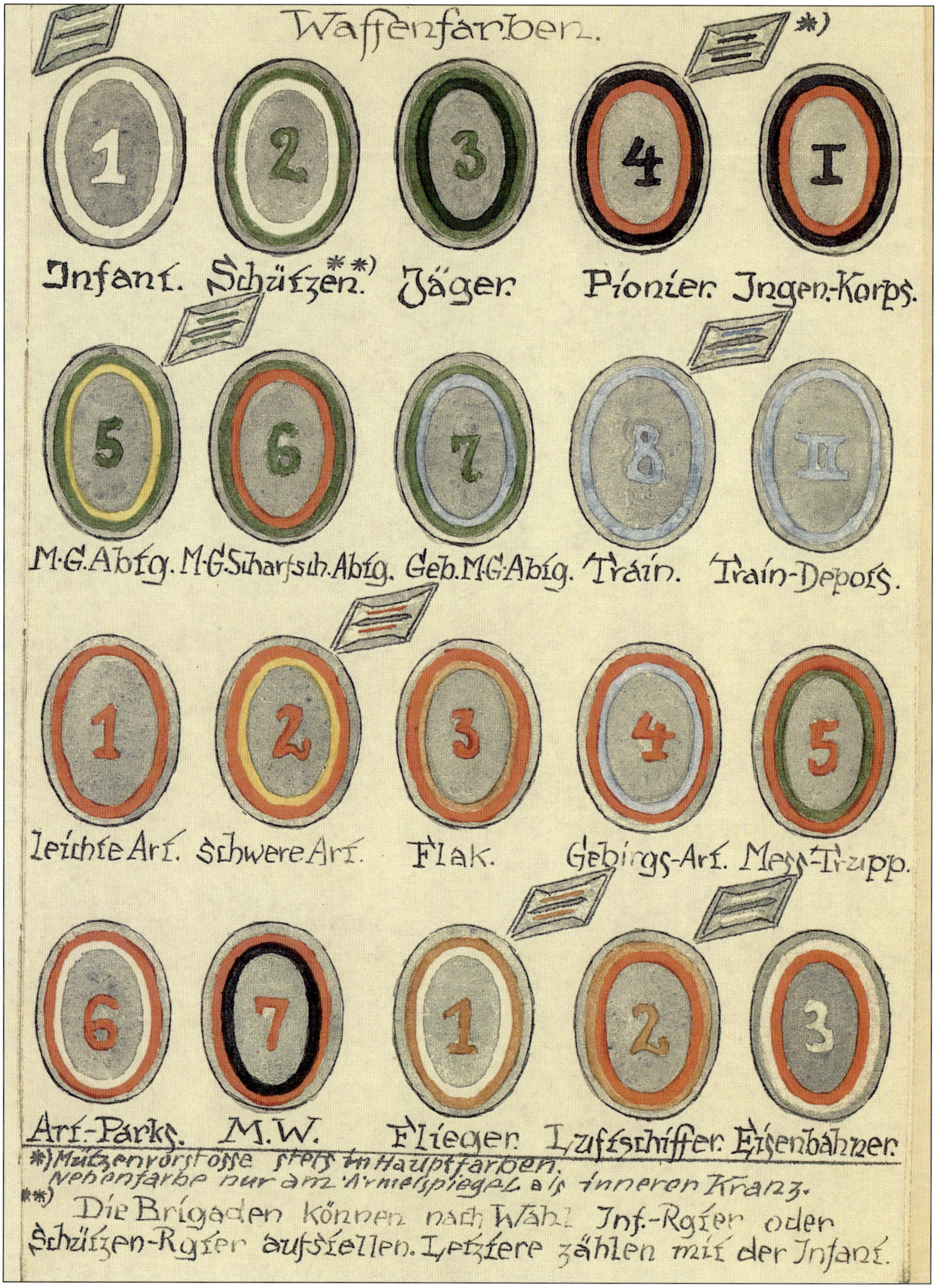

Uniform regulations for the early Reichswehr were published periodically starting in 1919 as the *Heers-Verordnungs-blat.Herausgegeben vom Reichswehrministerium (Heeres-verwalting).* This publication had a very small production run, plus a limited distribution, therefore, any use of color in the early stages was done by hand. The following three illustrations of Armspiegel are of that nature. *From the Anne S. K. Brown Military Collection, Brown University. Providence, Rhode Island.*

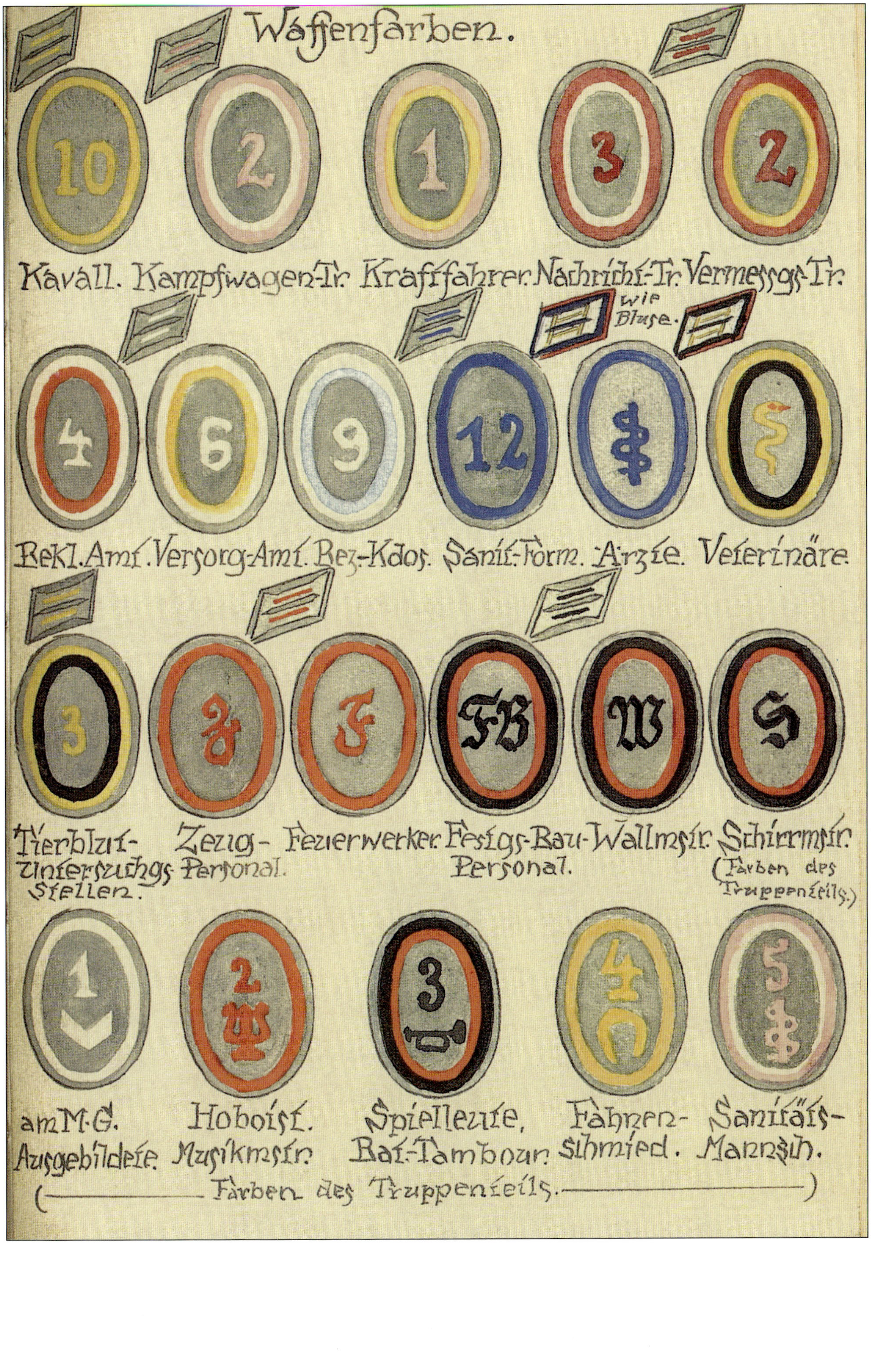
Waffenfarben.
Kavall.
Kampfwagen-Tr.
Kraftfahrer.
Nachricht-Tr.
Vermessgs-Tr.
wie Bluse.
Bekl.Amt.
Versorg-Amt.
Bez.-Kdos.
Sanit.-Form.
Ärzte.
Veterinäre.
Tierblut-untersuchgs-Stellen.
Zeug-Personal.
Feuerwerker.
Festgs-Bau-Personal.
Wallmstr.
Schirrmstr.
(Farben des Truppenteils.)
am M.G. Ausgebildete.
Hoboist. Musikmstr.
Spielleute, Bat-Tambour.
Fahnen-schmied.
Sanitäts-Mannsch.
(Farben des Truppenteils.)

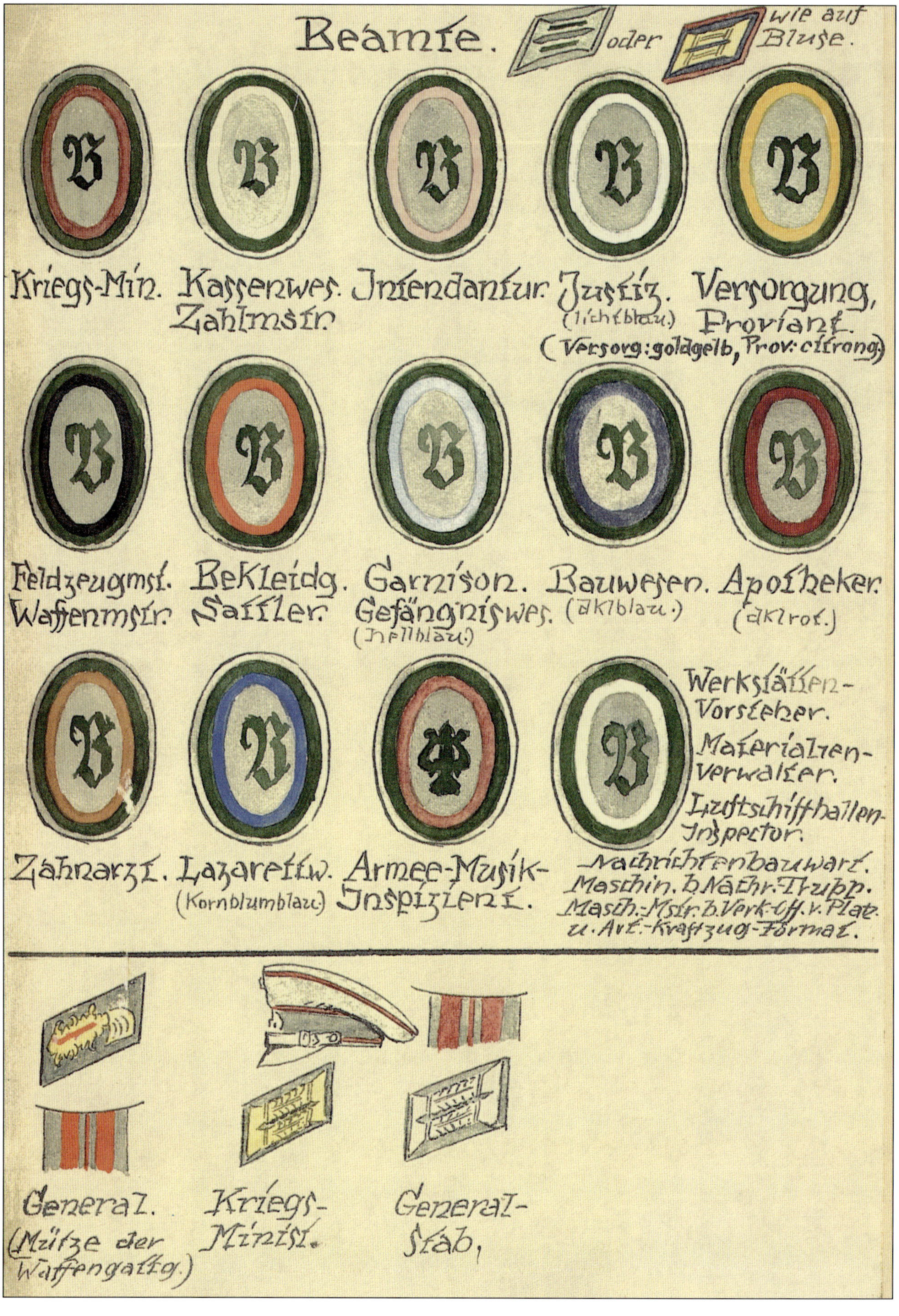
Beamte.
oder
wie auf Bluse.
Kriegs-Min.
Kassenwes. Zahlmstr.
Intendantur.
Justiz. (lichtblau)
Versorgung, Proviant.
(Versorg: goldgelb, Prov: citrong.)
Feldzeugmst. Waffenmstr.
Bekleidg. Sattler.
Garnison. Gefängniswes. (hellblau)
Bauwesen. (dklblau.)
Apotheker. (dklrot.)
Werkstätten-Vorsteher.
Materialien-verwalter.
Luftschiffhallen-Inspector.
Zahnarzt.
Lazarettw. (Kornblumblau)
Armee-Musik-Inspizient.
General. (Mütze der Waffengattg.)
Kriegs-Minist.
General-Stab,

From the Alte Armee to the Reichswehr came Husaren-Regiment von Zieten (Brandenburgisches) Nr. 3, first formed in 1730, becoming Reiterregiment Nr. 3. The only vestige of its once flamboyant self was the Dienstmütze still in the old regimental colors worn by the Oberwachtmeister as illustrated by Knötel the younger.

Above: This photo of young Reichswehr recruits from Stube 38 shows uniforms worn in 1921. The training NCO, a vizefeldwebel, is in the 1921 model Feldrock. The man standing second from the left is wearing a fly-fronted "regulation" tunic with two breast pockets. The rest are in M 1915 style Bluse and visorless Mützen, or Kratzchen.

OPPOSITE

Top: Members of I.R. 19 at leisure during the summer of 1923. Card playing, pipe smoking, beer drinking and dog petting are the activities of the hour by this mixed uniformed group. Fatigue and service dress are present and two men still wear their Freikorps von Epp badges on the left sleeve. The Dienstmütze visor caps show the combination of cocard, plus a wreath surrounding a Reichswehr insignia. The Kratzchen has a single cocard, as does the Feldmütze in the center.

Bottom: The sign tells that this is the "Club of the Women Haters", the meaning possibly lost in translation. The shoulder straps are marked "19" and the Gefreiter reading the paper has moved his Freikorps von Epp badge into the "vee" formed by his rank stripes. The armed and helmeted guard looks very much like the character from the old *Saturday Night Live* television show.

1923

Club der
Weiberfeinde.
1924
4. Komp. I.R. 125

In 1921 Verlag von Moritz Ruhl in Leipzig published *Die Deutsche Reichswehr* with twenty-three pages of text and twelve colored folding plates describing the Reichswehr as it was dressed 23 March 1921. The text gave the unit breakdown of Divisions 1-7 and Kavallerie Divisions 1-3. The seven Divisions consisted of infantry and artillery stationed at Königsberg, Stettin, Berlin, Dresden, Stuttgart, Münster, and Munich in that order. The cavalry 1-3 were stationed at Frankfurt, Breslau, and Cassel. The following uniform figures shown in color are the same poses and artwork as used in the 1919 publication picturing the uniforms and insignia two years prior. The twelve updated plates with the original captions are included here in their entirety.

Kavallerie-Offizier
im Mantel mit Stahlhelm

Truppen-Generalstab
Offizier

Infanterie
Stabsoffizier
Artillerie-Offizier
Adjutant

Jäger-Unteroffizier

Kraftfahr-Abteilung Offizier

Fahr-Abteilung Offizier

Infanterist
Soldat im Mantel mit Stahlhelm
Kavallerie-Gefreiter

Artillerist im Mantel

Pionier mit Stahlhelm

VII

Die Farben der Waffengattungen und der Heeresbeamten.

Die Waffenfarben der Truppen.

Infanterie (weiß)

Kavallerie (goldgelb)

Pioniere (schwarz)

Kraftfahrtruppen (rosa)

Sanitätsabteilungen (kornblumenblau)

Jäger (hellgrün)

Artillerie und Zeugämter (hochrot)

Fahrtruppen (hellblau)

Nachrichtentruppen (hellbraun)

Veterinäroffiziere (karmesinrot)

Offiziere des Reichswehrministeriums, der Inspektion für Waffen und Gerät, Truppengeneralstäbe (karmesinrot)

Die Farben der Heeresbeamten.

Haupt-Farbe | Neben-Farbe

Reichswehrministerium, Waffenschulen, Wehrkreiskommandos usw.

Heeresanwaltschaft

Technische Beamte

Lazarettverwaltung

Intendantur

Zahlmeister

Apotheker

Garnisonverwaltung

Dienstgradabzeichen auf dem linken Oberarm des Waffenrocks und Mantels.

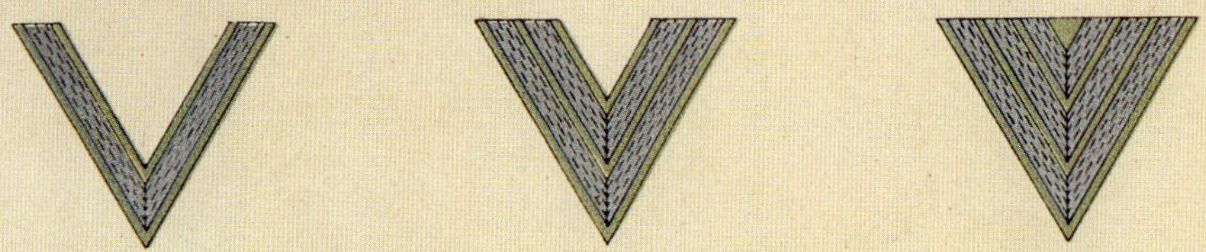

Oberschütze, Oberreiter, Oberpionier usw. — Gefreiter — Obergefreiter

Dienstgradabzeichen am Kragen des Waffenrocks.

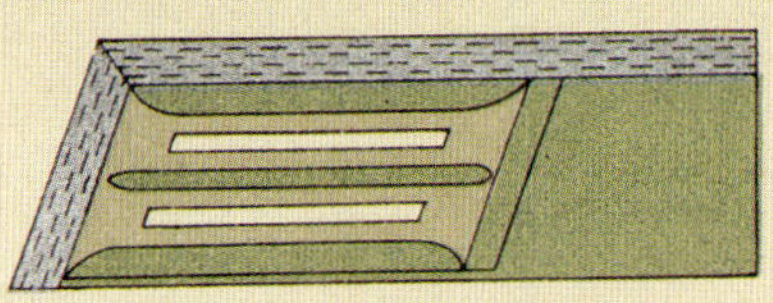

Tressenbesatz für alle Truppen (hier Infanterie)

Carl Starke, Kunstanstalt, Leipzig. — Moritz Ruhl, Verlag, Leipzig.

Nachdruck verboten! VIII

Dienstgradabzeichen an den Schulterklappen und Schulterstücken.

Schulterklappen mit Dienstgradabzeichen der Unteroffiziere.

Unteroffizier (Sanitätsabt.) — Unterfeldwebel, Fähnriche (Jäger) — Feldwebel, Wachtmeister, Oberfähnrich (Kavallerie) — Oberfeldwebel, Oberwachtmeister (Infanterie) — Schulterstücke für Musikmeister (Kavallerie) — Obermusikmeister (Infanterie)

Schulterklappen für Heeresbeamte.

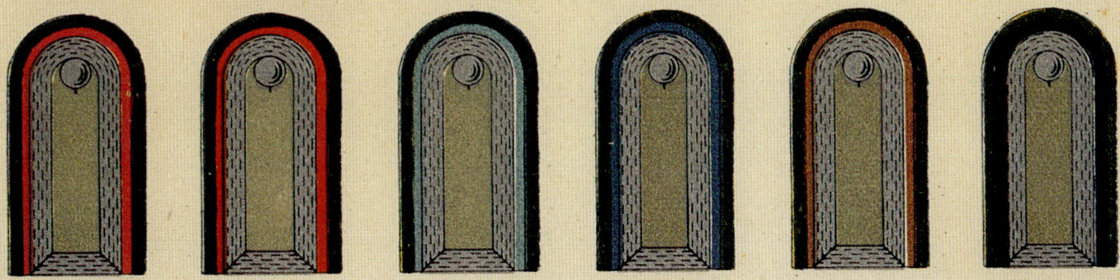

Reichswehr=ministerium usw. — Intendantur — Heeresanwaltschaft — Lazarettverwaltung — Garnisonverwaltung — Technische Beamte
(hier mit dem Range der Unterfeldwebel)

Schulterstücke der Offiziere.

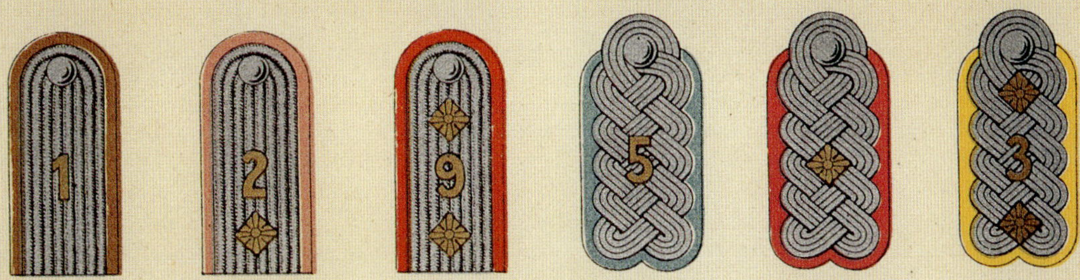

Leutnant (Nachrichtentruppen) — Oberleutnant (Kraftfahrtruppen) — Hauptmann, Rittmeister (Artillerie) — Major (Fahrtruppen) — Oberstleutnant (Generalstab) — Oberst (Kavallerie)

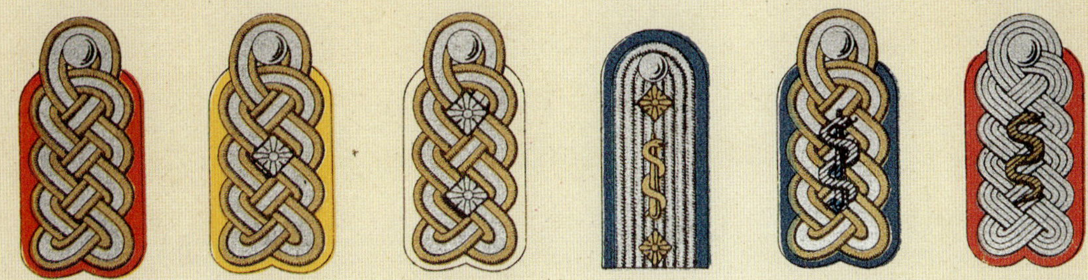

Generalmajor der Artillerie — Generalleutnant der Kavallerie — General der Infanterie — Stabsarzt — Generalstabsarzt — Oberstabsveterinär
Sanitäts- und Veterinär-Offiziere

Carl Starke, Kunstanstalt, Leipzig.

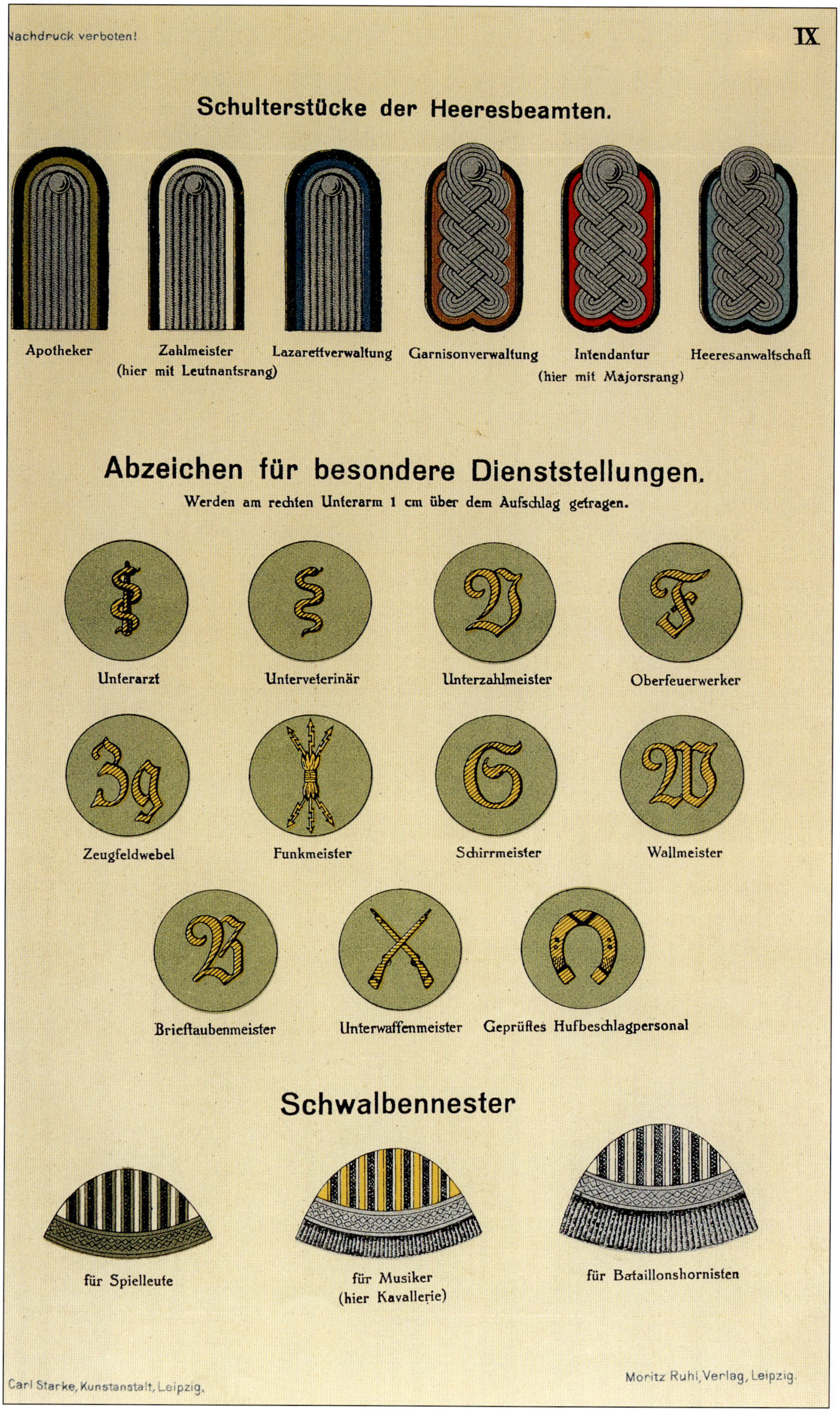
Nachdruck verboten!
IX
Schulterstücke der Heeresbeamten.
Apotheker
Zahlmeister (hier mit Leutnantsrang)
Lazarettverwaltung
Garnisonverwaltung
Intendantur (hier mit Majorsrang)
Heeresanwaltschaft
Abzeichen für besondere Dienststellungen.
Werden am rechten Unterarm 1 cm über dem Aufschlag getragen.
Unterarzt
Unterveterinär
Unterzahlmeister
Oberfeuerwerker
Zeugfeldwebel
Funkmeister
Schirrmeister
Wallmeister
Brieftaubenmeister
Unterwaffenmeister
Geprüftes Hufbeschlagpersonal
Schwalbennester
für Spielleute
für Musiker (hier Kavallerie)
für Bataillonshornisten
Carl Starke, Kunstanstalt, Leipzig.
Moritz Ruhl, Verlag, Leipzig.

Nachdruck verboten!
Kragen-Stickereien und Kragen-Litzen.
Stickereien am Waffenrock-Kragen der Offiziere
für Generale
für Offiziere des Reichswehrministeriums usw.
für Offiziere der Truppengeneralstäbe
Litzen am Waffenrock-Kragen der Truppen.
Infanterie
Jäger
Kavallerie
Artillerie
Pioniere
Fahrtruppen
Kraftfahrtruppen
Nachrichtentruppen
Sanitätsabteilungen
Veterinäre
Litzen am Waffenrock-Kragen der Heeresbeamten.
Reichswehrministerium
Intendantur
Heeresanwaltschaft
Zahlmeister
Technische Beamte
Apotheker
Lazarettverwaltung
Garnisonverwaltung
Carl Starke, Kunstanstalt, Leipzig.
Moritz Ruhl, Verlag, Leipzig

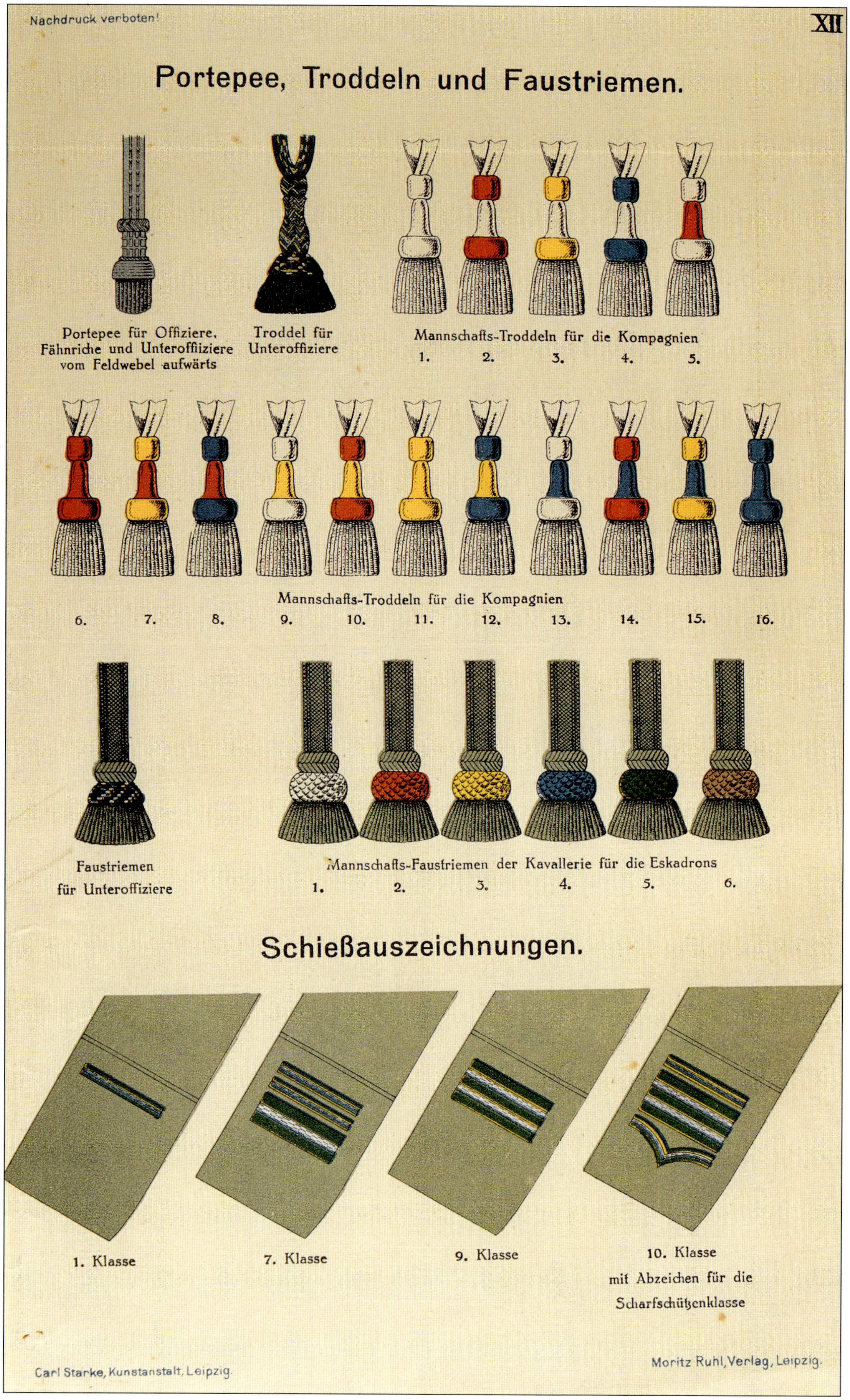
Nachdruck verboten!
XII
Portepee, Troddeln und Faustriemen.
Portepee für Offiziere, Fähnriche und Unteroffiziere vom Feldwebel aufwärts
Troddel für Unteroffiziere
Mannschafts-Troddeln für die Kompagnien
1. 2. 3. 4. 5.
Mannschafts-Troddeln für die Kompagnien
6. 7. 8. 9. 10. 11. 12. 13. 14. 15. 16.
Faustriemen für Unteroffiziere
Mannschafts-Faustriemen der Kavallerie für die Eskadrons
1. 2. 3. 4. 5. 6.
Schießauszeichnungen.
1. Klasse
7. Klasse
9. Klasse
10. Klasse mit Abzeichen für die Scharfschützenklasse
Carl Starke, Kunstanstalt, Leipzig.
Moritz Ruhl, Verlag, Leipzig.

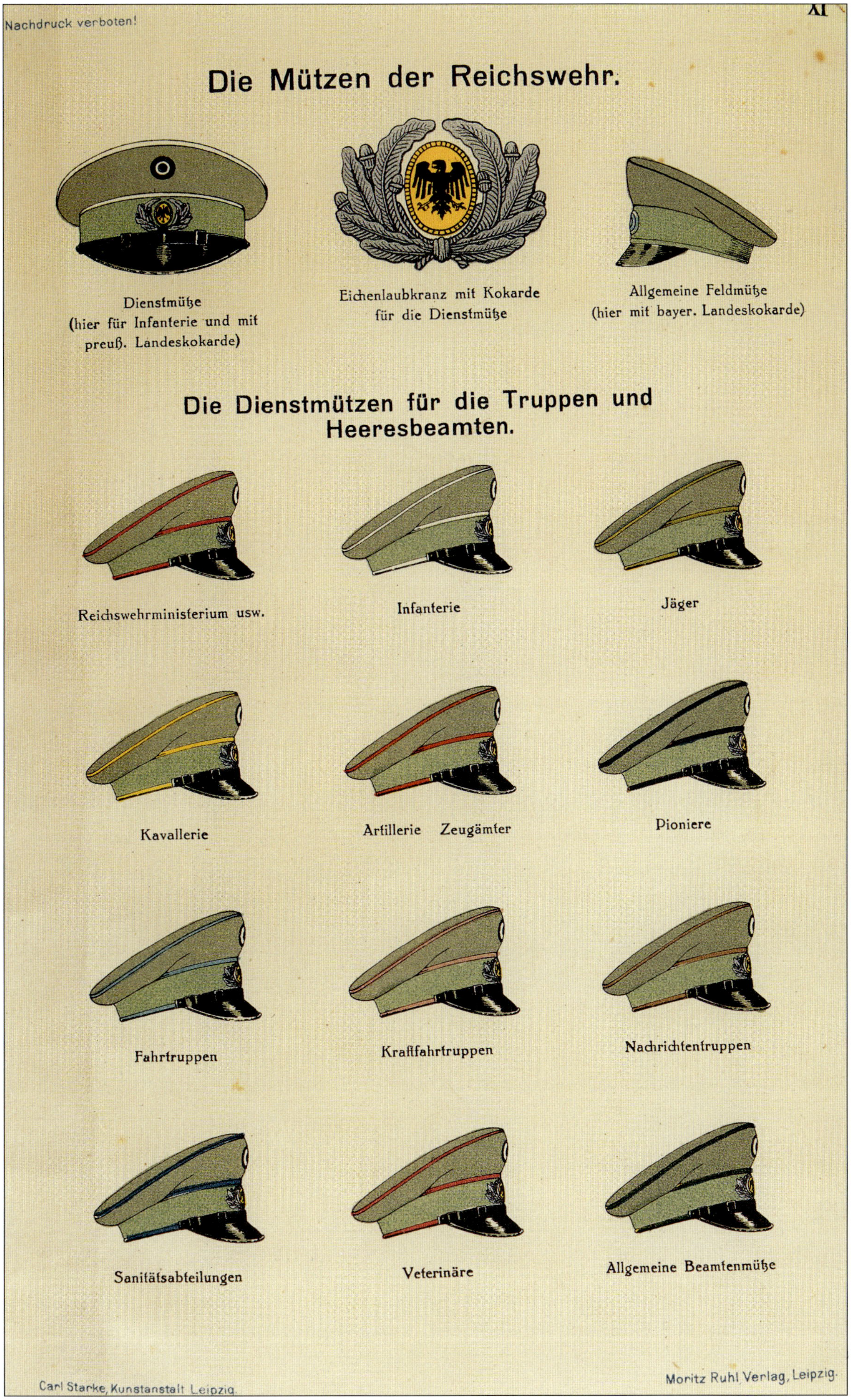
Nachdruck verboten!
XI
Die Mützen der Reichswehr.
Dienstmütze
(hier für Infanterie und mit
preuß. Landeskokarde)
Eichenlaubkranz mit Kokarde
für die Dienstmütze
Allgemeine Feldmütze
(hier mit bayer. Landeskokarde)
Die Dienstmützen für die Truppen und
Heeresbeamten.
Reichswehrministerium usw.
Infanterie
Jäger
Kavallerie
Artillerie Zeugämter
Pioniere
Fahrtruppen
Kraftfahrtruppen
Nachrichtentruppen
Sanitätsabteilungen
Veterinäre
Allgemeine Beamtenmütze
Carl Starke, Kunstanstalt Leipzig.
Moritz Ruhl Verlag, Leipzig.

Bibliography

Der Deutsche Militarismus. 2 volumes. Peter Bachmann, Kurt Zeisler. pp 908. Profusely illustrated. Militäverlag der Deutschen Demokzatischen Republik, Berlin 1971. German military history as presented by the Communist writers and publishers in the DDR in the 1970s.

The Nemesis of Power, The German Army in Politics 1918-1945. John W. Wheeler-Bennett, C.M.G., O.B.E. pp 829. 36 illustrations of Generals. St. Martin's Press, 1954

Deutsche Verbände und Truppen, 1918-1939. Altes Heer, Freiwilligen Verbände, Reichswehr, Heer, Luftwaffe, Landespolizei. Georg Tessin. pp 468. No illustrations. Biblio Verlag, Osnabrück, 1974. Listing of all old Army regiments in 1914, Freikorps and Volunteers 1919-20, Preliminary Reichsheer (100,000 man) 1921-1933, the Army, Luftwaffe, and the National Police 1933-39. A fine reference source of smaller Freikorps unit names and to which larger unit they belonged.

Fur das Stolze Edelweiss, Bildsband zur Geschichte von Freikorps Oberland und Bund Oberland. Peter Schuster. pp 190. Numerous black and white photo illustrations. Brienna Verlags GmbH., Aschau, 1999.

Die deutsche Reichswehr. Verlag Moritz Ruhl, Leipzig, 1921. pp 23. 12 colored folding plates included in this volume.

Die Uniformen des Friedensheers, der vorläufigen Reichswehr und des Reichsheers von 1919 bis 1921. Ingo G. Haarcke. Verlag Militaria-Archiv Klaus D. Potzwall. Hamburg, 1983. pp 62. 37 pages containing black and white illustrations.

Die Uniformierung und Ausrüstung des deutschen Reichsheers 1919-1932. Adolf Schlicht and Jürgen Kraus. Bayerischen Armeemuseums. Band 4, Ingolstadt 1987. pp 279. Black and white and color illustrations. One of the best books on the Reichswehr 1919-1932.

War, Mutiny and Revolution in the German Navy, the World War I Diary of Seaman Richard Stumpf. Edited and translated and with an Introduction by Daniel Horn. pp 442. 17 black and white illustrations, 1 map. Rutgers University Press, New Brunswick, New Jersey, 1967. A most interesting account from a sailor's point of view of service life and the Naval mutinies.

Deutsche Freikorps 1918-1921, Spezialkatalog. Lothar Hartung. pp 215. 205 pages of illustrations of Freikorpsbadges and portraits of Freikorps leaders. A price guide to value included. Published by the author, Junkers Hof, Germany.

The German Freikorps 1918-23. Carlos Cabellero Jurado. Illustrated by Ramiro Bajeiro. pp 84. Illustrated in color and black and white. Osprey Elite series, Oxford, U.K, 2001.

Freikorps Insignia. VerKuilen Ager. pp 67. Profusely illustrated with black and white drawings and photographs. Published by the author. Rochester, NY, 1979.

Handbuch Zur Geschichte Der Deutschen Freikorps. Robert Thoms, Stefan Pochanke. pp 244. 28 black and white illustrations. Published by MTM Verlag GmbH., 2001. Useful information included is a 679 entry list of Freikorps units, all sizes, dates of formation, if known, and type of unit down to company level. Short biography of Freikorpsführer, and an extensive bibliography.

Vangard of Nazism, The Free Corps Movement in Postwar Germany 1918-1923. Robert G. L. Waite. 344 pp. No illustrations. Harvard University Press, Cambridge, Massachusetts, 1952. Done as a doctoral thesis, this is one of the most scholarly subjects on the Freikorps.

Hitler's Heralds, The Story of the Freikorps 1918-1923. Nigel H. Jones. pp 284. 20 black and white illustrations. Dorset Press, New York, 1992. Not as scholarly as Waite, but a good readable account of the Freikorps history.

The Kings Depart, The Tragedy of Germany: Versailles and the German Revolution. Richard M. Watt. pp 604. 22 black and white illustrations. Simon and Schuster, New York, 1968. Covers the period from the Armistice to the Versailles Treaty in detail.

Histoire de L'Armée Allemande 1918-1945. 2 Volumes. Volume I, *de L'Armée Imperiale a la Reichswehr 1918-1919.* Volume II, *de la Reichswehr a L'Armée Nationale 1919-1938.* J. Benoist-Méchin. pp 1081, 2 maps. Unfortunately for the English speaking reader, only Volume I has been translated from the French. Titled *History of the German Army Since the Armistice Vol. I, From the Imperial Army to the Reichswehr.* Published by Scientia A. G. Zürich, 1939.

Revolution und Fotografie, Berlin 1918/19. Many contributing authors. pp 302. Profusely illustrated with black and white photos. Verlag Dirk Nishen, Berlin, 1989.

Fotografie und Revolution, Munchen 1918/19. Rudolf Herz and Dirk Halfbrodt. pp 326. Profusely illustrated with black and white photos many by Heinrich Hoffmann, later photographer to Hitler. Verlag Dirk Nishen, Berlin, 1988.

The Weimar Republic through the Lens of the Press. Torsten Palmier and Hendrik Neubauer. pp 415. Profusely illustrated. Könemann Verlagsgesellshaft mbH, Cologne, 2000.

Söldner Ohne Sold, Die Deutschen Freikorps 1918-1923. Dominque Venner. pp 409. No illustrations. First published in Paris in French under the title of *Baltikum*, German edition. Gustav Lübbe GmbH, Bergisch Gladbach.

Assize of Arms, the Disarmament of Germany and Her Rearmament (1919-1939), J. H. Morgan. pp 357. 12 illustrations and an end paper of German Army Corps Districts prior to the Versailles Treaty. Oxford University Press, New York, 1946.

History of the German General Staff 1657-1945. Walter Goerlitz. pp 508. 31 black and white illustrations. A translation from *Der Deutsche Generalstab.* First American Edition Frederick A. Praeger Inc., New York, 1953.

The German Officer-Corps in Society and State 1650-1948. pp 414. No illustrations. Weidenfeld and Nicolson, London, 1965.

A History of Modern Germany 1840-1945. Hajo Holborn. pp 844, 13 maps. No illustrations. The concluding volume of a three volume history of Germany from the Reformation. Princeton University Press, Princeton NJ, 1969. Paperback, 1982.

Das Buch vom Freikorpskampfer, Ernst von Salomon. pp 498. Hundreds of photos. Wilhelm Limpert Verlag, Berlin, 1938. Short histories of all the Freikorps campaigns written primarily by participants.

Polizei Greist ein Bilddokumente der Schutzpolizei. Hans Roden. pp 272. 330 photographs from the Polizeibild Archiv. Covers the history of the Deutschen Schützpolizei from 1919-1934. Verlag von Breitkopf und Härtel, Leipzig, 1934.

UNIFORMS & EQUIPMENT OF THE IMPERIAL GERMAN ARMY 1900-1918: A Study in Period Photographs. *Charles Woolley.* This initial volume, of a continuing study, features formal studio portraits of pre-war dress and wartime uniforms of all arms, as well as photo postal cards taken in the field of various troops. Also included is a sixty page full-color uniform section reproduced from the rare 1914 plates by Major Arthur Schmidt of Infanterie Regiment Nr. 172. A must volume for the serious military historian, collector, and World War I re-enactor.
Size: 9" x 12" • over 500 b/w photographs and 50 color drawings • 376 pp.
ISBN: 0-7643-0935-8 • hard cover • $69.95

UNIFORMS & EQUIPMENT OF THE IMPERIAL GERMAN ARMY 1900-1918: A Study in Period Photographs - Air Service • Cavalry • Assault Troops • Signal Troops • Pickelhauben • Steel Helmets • Vehicles. Charles Woolley. This second volume in the series contains over 500 never before published photographic images of the Imperial German military forces. Each photograph and caption has been carefully researched, affording the reader much information not to be found elsewhere, plus the inclusion of a glossary and an annotated bibliography which make this volume essential for the serious military historian, collector, and World War I re-enactor.
Size: 9" x 12" • over 500 b/w photos, over ninety color profiles • 320 pp.
ISBN: 0-7643-1104-2 • hard cover • $69.95

THE KAISER'S ARMY IN COLOR: Uniforms of the Imperial German Army as Illustrated by Carl Becker 1890-1910. *Charles Woolley.* The author brings to the student of uniforms, as well as the art and military historian, the true glory of this brief, but most colorful time. The 104 never before published uniform plates are a treasure trove of color, action, and rarity which have only been viewed by a select few before this publication. The additional illustrations from books and the postcard art reproduced here well display the scope and depth of Becker's work.
Size: 9" x 12" • over 130 color plates • 152 pp.
ISBN: 0-7643-1173-5 • hard cover • $49.95

HELMETS OF THE FIRST WORLD WAR: Germany, Britain & their Allies. *Michael J. Haselgrove & Branislav Radovic.* Superb color photographs, including multiple full-views and detail shots, depict over 150 helmets of Germany, Britain, France, United States, Austria, Turkey, and others from World War I. Previously unpublished World War I photographs show the helmets as they were worn.
Size: 9" x 12" • over 500 color and b/w photographs • 240 pp.
ISBN: 0-7643-1020-8 • hard cover • $75.00

GERMAN MILITARY RIFLES & MACHINE PISTOLS 1871-1945. *Hans-Dieter Gotz.* Richly illustrated volume covers the development of modern German rifles and machine pistols, as well as their ammunition, and includes many rare and experimental types. Covered are the Werder rifle, Mauser rifles, the various M/71 rifles and ammunition, the 88 cartridge, the Infantry Rifle 88, the 98 rifles, the Fallschirmjäger rifle, the 41 & 43 rifles, ERMA and Walther machine pistols and many more.
Size: 8 1/2" x 11" • over 200 photos, many documents/drawings • 248 pp.
ISBN: 0-88740-264-X • hard cover • $35.00

CAMOUFLAGE UNIFORMS OF THE WAFFEN-SS: A Photographic Reference. *Michael Beaver.* Finally a book that unveils the shroud of mystery surrounding Waffen-SS camouflage clothing. Illustrated here, both in full color and in contemporary black and white photographs, this unparalleled look at Waffen-SS combat troops and their camouflage clothing will benefit both the historian and collector.
Size: 8 1/2" x 11" • over 1,000 color and b/w photos • 296 pp.
ISBN: 0-88740-803-6 • hard cover • $69.95

GERMAN MILITARY RIBBON BARS 1914-1957. *Mark Hayden.* This book fills an obvious gap in reference works on German awards and medals. Drawing on many resources from around the world, his own collection, and with many unpublished photographs from 1914-1970, author Mark Hayden examines German ribbon bars from World War I, the Inter-War years, World War II and finally those worn by veterans in the reconstituted Bundeswehr of 1957. The wide variety of construction, styles, and combinations are discussed in detail, along with information on values, availability, and reproductions.
Size: 6"x9" • over 80 color and b/w photos • 72 pp.
ISBN: 0-7643-1206-5 • soft cover • $19.95

GEAR UP! Flight Clothing & Equipment of USAAF Airmen in WWII. *Jon A. Maguire.* Provides an in-depth examination of personal flight clothing and equipment used by United States Army Air Force flyers in World War II. In addition to U.S. issue items, the book also covers R.A.F. gear used by American airmen. The original items are presented in full color photographs, including detail shots and labels. Also included are hundreds of World War II era photographs showing the items in use, as they were worn. Portions of actual USAAF instructional manuals are provided as well. A valuable reference to historians, collectors, and modelers.
Size: 8 1/2" x 11" • over 600 b/w and color photos • 184 pp.
ISBN: 0-88740-744-7 • hard cover • $45.00

EAGLES RECALLED: Pilot and Aircrew Wings of Canada, Great Britain and the British Commonwealth 1913-1945. *Warren Carroll.* These pieces of cloth and metal symbolize the daring, bravery, suffering and loss of men who flew in deadly aerial battles for democratic freedom. Much of the material contained in this publication has never been published. The author has also made new historical discoveries presented here for the very first time - he has accessed private collections, photographed rare museum acquisitions, and received support from historians in seven countries over a period of some ten years. This work brings to readers a detailed and comprehensive study of the brevets issued to aviators who fought with Great Britain in World Wars I and II.
Size: 8 1/2" x 11" • over 800 color and b/w photos • 240 pp.
ISBN: 0-7643-0244-2 • hard cover • $79.95

VINTAGE FLYING HELMETS: Aviation Headgear Before The Jet Age. *Mick J. Prodger.* This new book explores the history and development of early flying headgear throughout the pioneering nations in military and sport aviation, from the earliest exploits of the Wright Brothers, to the end of World War II. Drawn from aviation museums and private collections from around the world, the illustrated color section features over 1000 photographs depicting more than 650 different items of flying headgear: helmets, goggles, oxygen masks and accessories, including many unique personalizations and rarely seen items.
Size: 9" x 12" • over 1,000 color and b/w photographs • 336 pp.
ISBN: 0-88740-776-5 • hard cover • $75.00

JET AGE FLIGHT HELMETS: Aviation Headgear in the Modern Age. *Alan R. Wise & Michael S. Breuninger.* This is the first book ever to cover in detail the history and development of military flight helmets from the post-World War II era to the present, and includes over 120 different helmets and their associated equipment such as oxygen masks, boom microphones, inner helmets etc. Specific details of each helmet include manufacturer, proper designation, unique features, accessories, periods of use, branch of service(s), and aircraft in which is was used-selected export users are also included. Mike Breuninger is also the author of *United States Combat Aircrew Survival Equipment World War II to the Present: A Reference Guide for Collectors.*
Size: 9" x 12" • over 1,000 color and b/w photographs • 248 pp.
ISBN: 0-7643-0070-9 • hard cover • $75.00